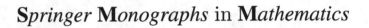

Springer **M**onographs in **M**athematics

For further volumes:
http://www.springer.com/series/3733

Steven Givant

Duality Theories for Boolean Algebras with Operators

 Springer

Steven Givant
Department of Mathematics
Mills College
Oakland, CA, USA

ISSN 1439-7382   ISSN 2196-9922 (electronic)
Springer Monographs in Mathematics
ISBN 978-3-319-35026-4   ISBN 978-3-319-06743-8 (eBook)
DOI 10.1007/978-3-319-06743-8
Springer Cham Heidelberg New York Dordrecht London

Mathematics Subject Classification (2010): 03G05, 03G25, 03B45, 03B53, 06E15, 06E25, 08C20, 18A20, 18B30

Contents

Introduction

There are two natural dualities for Boolean algebras. The first duality, sometimes called the *discrete duality*, is algebraic in nature; it concerns the duality between the category of all sets with mappings between sets as the morphisms, and the category of all complete and atomic Boolean algebras with complete homomorphisms between such algebras as the morphisms. The second duality—the famous Stone duality—is topological in nature and much deeper; it concerns the duality between the category of all Boolean spaces with continuous mappings between such spaces as the morphisms, and the category of all Boolean algebras with homomorphisms between such algebras as the morphisms. Each of these two dualities can be extended to Boolean algebras with normal operators, and in both cases the extensions are non-trivial, illuminating, and have important applications. There is a third duality for Boolean algebras with normal operators—a hybrid between the algebraic and the topological dualities—that seems not to have been notice before, even in the case of Boolean algebras, and although it may seem somewhat less natural than the other two dualities, it has important applications.

In this monograph, we develop these three dualities for Boolean algebras with *normal* operators (from now on called simply *Boolean algebras with operators*, for brevity, the normality of the operators being tacitly assumed).

The first chapter is concerned with algebraic duality. It begins by reviewing the known duality between arbitrary relational structures and arbitrary complete and atomic Boolean algebras with operators—or, what amounts to the same thing, the duality between relational structures and their complex algebras (up to isomorphism, complete and atomic Boolean algebra with operators are just complex algebra of relational structures—see Theorem 1.3 below). This duality dates back to Jónsson-Tarski [21].

The duality between structures is accompanied by a corresponding duality between morphisms. For relational structures $\mathfrak{U}$ and $\mathfrak{V}$ with corresponding dual complex algebras $\mathfrak{Cm}(U)$ and $\mathfrak{Cm}(V)$, every bounded homomorphism from $\mathfrak{U}$ to $\mathfrak{V}$ has a dual that is a complete homomorphism from $\mathfrak{Cm}(V)$ to $\mathfrak{Cm}(U)$, and conversely, every complete homomorphism from $\mathfrak{Cm}(V)$ to $\mathfrak{Cm}(U)$ has a dual that is a bounded homomorphism from $\mathfrak{U}$ to $\mathfrak{V}$. The dual of an epimorphism is a monomorphism and conversely. Furthermore, if the notion of dual morphism is defined carefully, then every bounded homomorphism between relational structures is its own second dual, as is every complete homomorphism between complex algebras. In particular, the correspondence between bounded homomorphisms from $\mathfrak{U}$ to $\mathfrak{V}$ and complete homomorphisms from $\mathfrak{Cm}(V)$ to $\mathfrak{Cm}(U)$ is a (bijective) contravariant functor that maps monomorphisms to epimorphisms, and epimorphisms to monomorphisms. For the special case of frames and modal algebras—that is to say, relational structures with a single binary relation, and complete and atomic Boolean algebras with a single complete unary operator—this duality of morphisms is stated in Thomason [39] with a brief hint of the proof. For the general case of arbitrary relational structures and arbitrary complete and atomic Boolean algebras with complete operators, one half of the duality of morphisms is given in Goldblatt [13] (and repeated in Blackburn-de Rijke-Venema [1] with a credit to Goldblatt). The other half is implicit in Jónsson [19] (and repeated explicitly in [1]), but no proof is provided; we provide a proof. Some related results are formulated at the end of Hansoul [16].

The duality between morphisms is exploited to establish, first of all, a duality between inner subuniverses of a relational structure $\mathfrak{U}$ and complete ideals in the complex algebra of $\mathfrak{U}$; and second of all, a duality between bounded congruence relations on $\mathfrak{U}$ and complete subuniverses of $\mathfrak{Cm}(U)$. One aspect of these dualities is a dual lattice isomorphism from the lattice of inner subuniverse of $\mathfrak{U}$ to the lattice of complete ideals in $\mathfrak{Cm}(U)$, and a dual lattice isomorphism from the lattice of bounded congruence relations on $\mathfrak{U}$ to the lattice of complete inner subuniverses of $\mathfrak{Cm}(U)$. These dualities imply corresponding dualities between the inner substructures of $\mathfrak{U}$ and the complete quotients of $\mathfrak{Cm}(U)$ on the one hand, and between bounded quotients of $\mathfrak{U}$ and complete subalgebras of $\mathfrak{Cm}(U)$ on the other hand. For example, the dual complex algebra of each inner substructure $\mathfrak{V}$ of $\mathfrak{U}$ is isomorphic to the quotient of $\mathfrak{Cm}(U)$ modulo the complete ideal that is the dual of the universe of $\mathfrak{V}$, and vice versa. Similarly, the dual complex algebra of each quotient of $\mathfrak{U}$ modulo a bounded congruence Θ is isomorphic to the complete subalgebra of $\mathfrak{Cm}(U)$ whose universe is the dual of the congruence Θ, and vice versa.

In the final part of Chapter 1, sharper forms are established for two theorems in Goldblatt [13] (see also Thomason [39] and Goldblatt [12] for the special case of frames and modal algebras). First of all, it is shown in [13] that the complex algebra of the disjoint union of a system of relational structures is isomorphic to the (external, or Cartesian) direct product of the corresponding system of complex algebras. The result is strengthened here to show that the complex algebra of the disjoint union is actually equal to (and not just isomorphic to) the *internal* direct product of the system of complex algebras. This strengthened form plays an important role in the next chapter, on topological duality. Secondly, it is shown in [13] (generalizing a result of Monk [26]) that an ultraproduct of a system of complex algebras modulo an ultrafilter D is embeddable into the complex algebra of the ultraproduct of the corresponding system of dual relational structures modulo D. This result is strengthened here to show that the complex algebra of the ultraproduct of relational structures is, up to isomorphism, just the completion of the ultraproduct of the system of complex algebras. In other words, the algebraic dual of an ultraproduct of a system of relational structures is just the completion of the corresponding ultraproduct of the system of dual complex algebras.

The second chapter is concerned with topological duality. It studies the duality between arbitrary Boolean algebras with operators (not just complete and atomic algebras) and arbitrary relational spaces, that is to say, arbitrary relational structures endowed with the topology of a Boolean space under which the fundamental relations of the structure are continuous and clopen. This topological duality was first investigated by Halmos [15] and Goldblatt [11, 12] for the case of Boolean algebras with a single unary operator, and by Goldblatt [13] for arbitrary Boolean algebras with operators. (Both authors build on early work of Jónsson-Tarski [21], but Goldblatt is in fact concerned with the duality between bounded distributive lattices with operators and Priestley spaces, so he also builds on the work of Priestley [30]. The reports [12] are published versions of Goldblatt's doctoral dissertation [11], so we shall always refer to them instead of to [11].)

Hansoul [16] and Sambin-Vaccaro [32] contain related developments. Our approach is different from that of Goldblatt. We show, however, that our approach (which seems closer to the standard treatment of topological algebraic structures) is equivalent to his approach.

The duality between structures is accompanied by a corresponding duality between morphisms. For Boolean algebras with operators $\mathfrak{A}$ and $\mathfrak{B}$, and corresponding dual relational spaces $\mathfrak{U}$ and $\mathfrak{V}$, every homomorphism (not necessarily complete) from $\mathfrak{A}$ to $\mathfrak{B}$ has a dual that is a continuous bounded homomorphism from $\mathfrak{V}$ to $\mathfrak{U}$, and conversely, every continuous bounded homomorphism from $\mathfrak{V}$ to $\mathfrak{U}$ has a dual that is a homomorphism from $\mathfrak{A}$ to $\mathfrak{B}$. The dual of an epimorphism is a monomorphism and conversely. Furthermore, if the notion of dual morphism is defined carefully, then every homomorphism between Boolean algebras with operators is its own second dual, as is every continuous bounded homomorphism between relational spaces. In particular, the correspondence between homomorphisms from $\mathfrak{A}$ to $\mathfrak{B}$ and continuous bounded homomorphisms from $\mathfrak{V}$ to $\mathfrak{U}$ is a (bijective) contravariant functor that maps monomorphisms to epimorphisms, and epimorphisms to monomorphisms. For the special case of Boolean algebras with a single unary operator and relational spaces with a single binary relation, versions of this duality of morphisms are stated in Halmos [15] and Goldblatt [12]. A general version of the duality of morphisms for arbitrary relational spaces and arbitrary Boolean algebras with operators is given in Goldblatt [13] (and repeated in [1] with a credit to Goldblatt). See also Hansoul [16] and Sambin-Vaccaro [32] for related developments.

This duality is exploited here to show that there is, first of all, a duality between ideals (not necessarily complete) in an arbitrary Boolean algebra with operators $\mathfrak{A}$ and special open subsets of the dual relational space $\mathfrak{U}$; and second of all, a duality between subuniverses (not necessarily complete) of $\mathfrak{A}$ and (bounded) relational congruences on $\mathfrak{U}$. One aspect of these dualities is a lattice isomorphism from the lattice of ideals in $\mathfrak{A}$ to the lattice of special open subsets of $\mathfrak{U}$, and a dual lattice isomorphism from the lattice of subuniverses of $\mathfrak{A}$ to the lattice of relational congruences on $\mathfrak{U}$. These dualities imply corresponding dualities between the structures themselves. For example, if $\mathfrak{U}$ is a relational space and $\mathfrak{A}$ the dual Boolean algebra with operators, then the dual relational space of the quotient of $\mathfrak{A}$ modulo an ideal is, up to homeo-isomorphism, the inner subspace of $\mathfrak{U}$ whose universe is the complement of the special open set that is the dual of the ideal; and vice versa. Similarly, the dual relational space of a subalgebra $\mathfrak{B}$ of $\mathfrak{A}$ is, up to homeo-isomorphism, the quotient of $\mathfrak{U}$ modulo the relational congruence that is the dual of the universe of $\mathfrak{B}$.

Next, we take up the problem of describing the dual relational space $\mathfrak{U}$ of a Boolean algebra with operators $\mathfrak{A}$ that satisfies some completeness condition. It is shown, for example, that $\mathfrak{A}$ is complete as an algebra if and only if $\mathfrak{U}$ is complete as a relational space. This result can be extended to weaker forms of completeness. For example, $\mathfrak{A}$ is countably complete as an algebra if and only if $\mathfrak{U}$ is countably complete as a relational space.

In the final part of Chapter 2, the duality between morphisms is used to describe the dual spaces of direct and subdirect products of systems of Boolean algebras with operators. Consider a disjoint system $(\mathfrak{U}_i : i \in I)$ of relational spaces, and suppose $\mathfrak{U}$ is the union of this system. Let $(\mathfrak{A}_i : i \in I)$ be the corresponding system of dual Boolean algebras with operators, and let $\mathfrak{A}$ be the internal direct product of this system, and $\mathfrak{D}$ the internal weak direct product. If the systems in question are finite (that is to say, if the index set I is finite), then the dual algebra of the relational space $\mathfrak{U}$ is just $\mathfrak{A}$; put another way, the dual of the direct product of finitely many Boolean algebras with operators is the disjoint union of the corresponding dual relational spaces. When the systems in question are infinite, the situation becomes more complicated to describe. The dual of every compactification of $\mathfrak{U}$ is isomorphic (via relativization) to a subalgebra of the direct product $\mathfrak{A}$, and in fact to a subalgebra that includes the weak direct product $\mathfrak{D}$; moreover, every subalgebra of $\mathfrak{A}$ that includes $\mathfrak{D}$ is the isomorphic image (via relativization) of some compactification of $\mathfrak{U}$. Compactifications of $\mathfrak{U}$ are mapped via this duality correspondence to the same subalgebra of $\mathfrak{A}$ if and only if the compactifications are equivalent in the sense that they are homeo-isomorphic over $\mathfrak{U}$, so one may speak with some justification of the dual Boolean algebra with operators of an equivalence class of compactifications of $\mathfrak{U}$. The function that maps each such equivalence class to the isomorphic copy (via relativization) of its dual Boolean algebra with operators is a lattice isomorphism from the lattice of equivalence classes of compactifications of $\mathfrak{U}$ to the lattice of subalgebras of the direct product $\mathfrak{A}$ that include the weak direct product $\mathfrak{D}$.

The preceding lattice isomorphism implies that the union space $\mathfrak{U}$ has a maximum compactification, and the dual Boolean algebra with operators of this maximum compactification is isomorphic (via relativization) to the direct product $\mathfrak{A}$. We prove that this maximum compactification of $\mathfrak{U}$ is just the Stone-Čech compactification of $\mathfrak{U}$. There are several different ways in which this result may be interpreted, and they all turn out to be true.

While the first chapter is concerned with the algebraic duality between relational structures and complete, atomic Boolean algebras with operators (that is to say, complex algebras of relational structures), and the second chapter is concerned with the topological duality between relational spaces and arbitrary Boolean algebras with operators, the third chapter is concerned with a hybrid duality that combines aspects of the algebraic and the topological dualities. Every weakly bounded homomorphism from a relational *structure* $\mathfrak{U}$ into a relational *space* $\mathfrak{V}$ has a dual homomorphism from the Boolean algebra with operators that is the topological dual of $\mathfrak{V}$ to the complete and atomic Boolean algebra with operators that is the algebraic dual of $\mathfrak{U}$—namely the complex algebra $\mathfrak{Cm}(U)$—and vice versa; and each of these morphisms is its own second dual. The epi-mono duality no longer holds: duals of epimorphisms are monomorphisms, but duals of monomorphisms may fail to be epimorphisms.

An arbitrary relational structure $\mathfrak{U}$ may be turned into a *locally compact* relational space (as opposed to a *compact* relational space, as considered in Chapter 2) by endowing $\mathfrak{U}$ with the discrete topology in which every subset of $\mathfrak{U}$ is simultaneously open and closed. We refer to such a discretely topologized relational structure as a *discrete space*. The hybrid duality mentioned above allows us to characterize the dual relational spaces (in the sense of Chapter 2) of all those subalgebras of $\mathfrak{Cm}(U)$ that contain the singleton subsets (or what amounts to the same thing, that contain the finite subsets) of $\mathfrak{U}$: they are just the weak compactifications of the discrete space $\mathfrak{U}$. Weak compactifications of $\mathfrak{U}$ are mapped via this duality correspondence to the same subalgebra of $\mathfrak{Cm}(U)$ if and only if the weak compactifications are equivalent in the sense that they are homeo-isomorphic over $\mathfrak{U}$, so one may speak with some justification of the dual Boolean algebra with operators of an equivalence class of weak compactifications of $\mathfrak{U}$. The function that maps each such equivalence class to the isomorphic copy (via relativization) of its dual Boolean algebra with operators is shown to be a lattice isomorphism from the lattice of equivalence classes of weak compactifications of $\mathfrak{U}$ to the lattice of subalgebras of $\mathfrak{Cm}(U)$ that contain the singleton subsets of $\mathfrak{U}$.

By restricting this result to the special case where there are no operators, we obtain the corollary—apparently new for Boolean algebras—that the function mapping each compactification of a discrete topological space U to the isomorphic copy (via relativization) of its dual Boolean algebra is an isomorphism from the lattice of equivalence classes of compactifications of U to the lattice of subalgebras of $Sb(U)$ (the Boolean algebra of subsets of U) that include the finite-cofinite subalgebra.

The lattice isomorphism mentioned above implies that there is a maximum weak compactification of a discrete space $\mathfrak{U}$, and the dual Boolean algebra with operators of this maximum weak compactification is isomorphic (via relativization) to the complex algebra $\mathfrak{Cm}(U)$. It is shown that this maximum weak compactification is in fact the Stone-Čech weak compactification of $\mathfrak{U}$. There are several ways in which this result may be interpreted, and all of them turn out to be true. This theorem may be viewed as a generalization of the well-known theorem in Boolean algebra that the dual of the Boolean algebra of all subsets of a set U is, up to homeomorphism, the Stone-Čech compactification of the discretely topologized space U.

The notions and results in this work not explicitly credited to others are due to the author. Many of them are extensions to Boolean algebras with operators of known notions and results for Boolean algebras (without operators). This applies in particular to the notions and results in Chapter 2 (see, for example, Chapters 34–38 and 43 in [10], or Chapter 3 in Koppelberg [23]). The notions and results in Chapter 3 appear to be new not only for Boolean algebras with operators, but also for Boolean algebras themselves. As mentioned above, some aspects of the duality theories discussed here have been extended to distributive lattices with operators. In this connection, the reader is referred to Goldblatt [13] and to the work of Mai Gehrke and her collaborators (see, for example, Gehrke [7], where further references to the literature may be found).

An understanding of the basic arithmetic and algebraic theory of Boolean algebras—say, along the lines developed in Chapters 6–8, 11, 12, 14, 17, 18, 20, and 26 of [10]—is assumed in this monograph. In particular, familiarity with fundamental laws of Boolean algebra and the algebraic notions of subuniverse, subalgebra, homomorphism, direct product, congruence, ideal, filter, and atom is helpful. In Chapter 2, some knowledge of basic topology (not too much) is also assumed. In particular, familiarity with the notions of open set, closed set, dense set, closure of a set, compact space, Hausdorff space, quotient topology, subspace topology, and continuous function is helpful. The necessary algebraic and topological notions and results (for the most part, with proofs) can all be found, for example, in [10]. We shall use that work as our standard reference on Boolean algebra and topology.

The author wishes to acknowledge his indebtedness to Professors Johan van Benthem, Robert Goldblatt, and Ian Hodkinson for reading through portions of the manuscript and providing him with encouragement and references to the literature; and to Professor Mai Gehrke for pointing out additional references to the literature.

Chapter 1
Algebraic Duality

In this chapter we study the algebraic duality that exists between relational structures and complete and atomic Boolean algebras with operators. The duality between the structures and algebras carries with it a corresponding duality between morphisms: every bounded homomorphism between relational structures corresponds to a dual complete homomorphism between the dual algebras, and conversely. The duality between the morphisms implies other dualities as well. Here are some examples. Every inner subuniverse of a relational structure corresponds to a complete ideal in the dual algebra, and vice versa. Every bounded congruence on a relational structure corresponds to a complete subuniverse of the dual algebra, and vice versa. The disjoint union of a system of relational structures corresponds to the direct product of the system of dual algebras, and vice versa.

1.1 Algebraic Duality for Boolean Algebras

In order to motivate the subsequent development, it is helpful to review some important aspects of the algebraic duality that exists between sets and complete, atomic Boolean algebras (without operators). The reader can find a more detailed presentation in [10].

Every set U is naturally correlated with a complete and atomic Boolean algebra, namely the Boolean algebra of all subsets of U. This algebra is denoted by $Sb(U)$ and is called the (*first*) *dual* of the set U. Inversely, every complete and atomic Boolean algebra A is naturally correlated with a set, namely the set of atoms in A. This set is called the (*first*) *dual* of the algebra A. If one starts with a set U, forms

S. Givant, *Duality Theories for Boolean Algebras with Operators*,
Springer Monographs in Mathematics, DOI 10.1007/978-3-319-06743-8_1,
© Springer International Publishing Switzerland 2014

its dual complete and atomic Boolean algebra A, and then forms the dual set of A, the result is the set V of all singletons of elements in U. Obviously, U and V are isomorphic (in the sense of being bijectively equivalent) via the mapping that takes each element in U to its singleton. Similarly, if one starts with a complete and atomic Boolean algebra A, forms its dual set U of atoms, and then forms the dual complete and atomic Boolean algebra of U, the result is a complete and atomic Boolean algebra B in which the elements are the subsets of the set of atoms in A. The function that maps each element r in A to the set of atoms that are below r is a Boolean isomorphism from A to B. This state of affairs is expressed by saying that each set and each complete and atomic Boolean algebra is isomorphic to its second dual. In practice, every set is identified with its second dual, and every complete and atomic Boolean algebra is also identified with its second dual. This means that every element u in a set U is identified with its singleton $\{u\}$, and every element r in a Boolean algebra A is identified with the set of atoms in A that are below r. Consequently, instead of speaking about U and its second dual V, and A and its second dual B, we may speak just of U and $B = Sb(U)$. This simplifies and clarifies the presentation quite a bit.

The duality between sets and complete and atomic Boolean algebras carries with it a corresponding duality between the functions on sets and the complete Boolean homomorphisms on the algebras. If ϑ is a mapping from a set U into a set V, then there is a natural mapping φ from $Sb(V)$ into $Sb(U)$ that is defined by

$$\varphi(X) = \vartheta^{-1}(X) = \{u \in U : \vartheta(u) \in X\}$$

for subsets X of V, and φ proves to be a complete Boolean homomorphism from $Sb(V)$ into $Sb(U)$. This complete homomorphism is called the (*first*) *dual* of the mapping ϑ. Inversely, if φ is a complete homomorphism from $Sb(V)$ into $Sb(U)$, then there is a natural mapping ϑ from U into V that is defined by

$$\vartheta(u) = r \qquad \text{if and only if} \qquad r \in \bigcap \{X \subseteq V : u \in \varphi(X)\},$$

or, equivalently,

$$\vartheta(u) = r \qquad \text{if and only if} \qquad u \in \varphi(\{r\}),$$

for elements r in U. This mapping is called the (*first*) *dual* of the complete homomorphism φ. If one starts with a complete homomorphism φ from $Sb(V)$ to $Sb(U)$, forms the dual mapping ϑ from U to V, and then forms the dual of ϑ, the result is the original complete homomorphism φ. Similarly, if one starts with a mapping ϑ from U to V, forms the dual complete homomorphism φ from $Sb(V)$ to $Sb(U)$, and then forms the dual of φ, the result is the original mapping ϑ. This state of affairs is expressed by saying that every mapping between sets U and V, and every complete homomorphism between the corresponding Boolean algebras $Sb(V)$ and $Sb(U)$ is its own second dual. A mapping ϑ from U to V is one-to-one or onto if and only if the dual complete homomorphism from $Sb(V)$ to $Sb(U)$ is onto or one-to-one respectively. Finally, if ϑ maps a set U to a set V, and δ maps V to a set W, and if φ and ψ are the respective duals of ϑ and δ, then the dual of the composition $\delta \circ \vartheta$ is just the composition $\varphi \circ \psi$. The category of sets with mappings as morphisms is therefore dually equivalent to the category of complete and atomic Boolean algebras with complete homomorphisms as morphisms.

1.2 Boolean Algebras with Operators

We begin our development by reviewing the basic definitions, notation, and terminology that we shall use. In general, we shall use the set-theoretical definition of a natural number n as the set of its predecessors,

$$n = \{0, \ldots, n-1\}.$$

Consequently, the phrase "for $i < n$" means "for $i = 0. \ldots, n-1$", and it is equivalent to the phrase "for $i \in n$".

A Boolean algebra is an algebra of the form

$$(A, +, -),$$

where A is a non-empty set, $+$ is a binary operation on A called *addition*, and $-$ is a unary operation on A called *complement*. Other common Boolean operations, distinguished constants, and relations are defined for A in the usual way. For instance, binary operations $\cdot$ of *multiplication* and $\ominus$ of *symmetric difference* are defined on A by

$$r \cdot s = -(-r + -s) \qquad \text{and} \qquad r \ominus s = (r \cdot -s) + (-r \cdot s)$$

for elements r and s in A, distinguished constants *zero* and *one* are defined in A by

$$0 = -(r + -r) \qquad \text{and} \qquad 1 = r + -r$$

(where r is an arbitrary element in A), and a *partial order* $\leq$ is defined on A by

$$r \leq s \qquad \text{if and only if} \qquad r + s = s.$$

The class of Boolean algebra may be axiomatized in several different ways. We assume that the reader is familiar with some axiomatization and with the basic laws of Boolean algebra that are a consequence of this axiomatization (see, for example, Chapters 2 and 6–8 in [10]). In general, we shall rarely cite specific Boolean laws in our proofs, but rather shall simply say that a certain conclusion follows "by Boolean algebra".

The *supremum* or *sum* of a set X of elements in a Boolean algebra is the least upper bound r of X in the sense that r is above every element in X (in the sense of the partial order defined above), and every other upper bound of X is above r. If the supremum of X exists, then we shall denote it by $\sum X$. Notice that the supremum of the empty set is 0, since 0 is obviously an upper bound, and in fact the least upper bound, of the empty set. Dually, the *infimum* or *product* of a set X of elements is the greatest lower bound s of X in the sense that s is a lower bound of X, and every other lower bound of X is below s. If the infimum of X exists, then we shall denote it by $\prod X$. Notice that the infimum of the empty set is 1, since 1 is obviously a lower bound, and in fact the greatest lower bound, of the empty set. A Boolean algebra is called *complete* if the supremum and infimum of every set of elements in the algebra exists. A necessary and sufficient condition for the algebra to be complete is that the supremum of every subset exists. An *atom* in the algebra is defined to be a minimal non-zero element (in the sense of the defined partial order), and the algebra is called *atomic* if every non-zero element is above an atom.

A *field of sets* is a Boolean algebra in which the universe consists of some (but not necessarily all) subsets of a set U, and the basic operations are the set-theoretic ones of forming unions and complements with respect to U. The set-theoretic operations of union, intersection, and complement are denoted respectively by $\cup$, $\cap$, and $\sim$.

An *ideal* in a Boolean algebra with universe A is a subset M of A that contains 0, that is *closed under addition* in the sense that $r + s$ belongs to M whenever r and s are both in M, and that contains $r \cdot s$ whenever r is in M and s in A. The first condition is equivalent to the condition that M be non-empty; the last condition is equivalent to the condition that M be *downward closed* in the sense that if r is in M

and if $s \leq r$, then s is in M. An ideal M is said to be *maximal* if it is a *proper* ideal—that is to say, if M is different from A—and if A is the only ideal that properly includes M. The *Maximal Ideal Theorem* for Boolean algebras says that every proper ideal can be extended to a maximal ideal (see, for example, Theorem 12 in [10]).

Dually, a *filter* in the Boolean algebra is a subset N of A that contains 1, that is *closed under multiplication* in the sense that $r \cdot s$ belongs to N whenever r and s are both in N, and that contains $r + s$ whenever r is in N and s in A. The first condition is equivalent to the condition that N be non-empty; the last condition is equivalent to the condition that N be *upward closed* in the sense that if r is in N and if $r \leq s$, then s is in N. A filter N is said to be *maximal* if it is proper and if the only filter in the algebra that properly includes N is the improper filter. A maximal Boolean filter is called an *ultrafilter*. A proper filter N is an ultrafilter just in case, for every element r in the algebra, N contains either r or $-r$. A subset X of the Boolean algebra has the *finite meet property* if the product of any finite number of elements in X is non-zero. For fields of sets, the terminology *finite intersection property* is often employed. As is well known, every subset with the finite meet property can be extended to an ultrafilter (see Exercise 12 in Chapter 20 of [10]).

For an ideal M in a Boolean algebra A, the set of complements

$$-M = \{-r : r \in M\}$$

is a filter in A, and conversely, for a filter N in A, the set of complements $-N$ is an ideal in A. The set $-M$ is called the *dual (Boolean) filter* of M, and $-N$ is called the *dual (Boolean) ideal* of N. An ideal or a filter is proper or maximal if and only if its dual is proper or maximal respectively. In fact, the function that maps each ideal to its dual filter is a lattice isomorphism from the lattice of ideals in A to the lattice of filters in A (see, for example, pp. 168–169 in [10]).

We turn now to the notion of a Boolean algebra with operators, and related notions, which were introduced and studied for the first time by Jónsson and Tarski in [21]. An operation f of *rank* n—that is to say, an operation of n arguments—on the universe A of a Boolean algebra is said to be *distributive* if it is distributive over addition in each argument in the sense that for each index $i < n$ and for each sequence

$$r_0, \ldots, r_{i-1}, r_{i+1}, \ldots, r_{n-1}, s, t$$

of elements in A,

$$f(r_0, \ldots, r_{i-1}, s + t, r_{i+1}, \ldots, r_{n-1})$$
$$= f(r_0, \ldots, r_{i-1}, s, r_{i+1}, \ldots, r_{n-1})$$
$$+ f(r_0, \ldots, r_{i-1}, t, r_{i+1}, \ldots, r_{n-1}).$$

For example, a binary operation $\circ$ on A is distributive if

$$r \circ (s + t) = (r \circ s) + (r \circ t) \qquad \text{and} \qquad (s + t) \circ r = (s \circ r) + (t \circ r)$$

for all r, s, and t in A. A distributive operation on A is called an *operator* (on A). Such an operator f always has the following *general distributivity property*: for any sequence $X_0, \ldots, X_{n-1}$ of finite, non-empty subsets of A, writing $t_i = \sum X_i$, we have

$$f(t_0, \ldots, t_{n-1}) = \sum \{f(r_0, \ldots, r_{n-1}) : r_i \in X_i \text{ for } i < n\}. \qquad (1)$$

The proof by induction is straightforward and is left to the reader (see Theorem 1.6 in [21]). An operator f of rank n is always *monotone* in the sense that for all sequences $r_0, \ldots, r_{n-1}$ and $s_0, \ldots, s_{n-1}$ of elements in A, if $r_i \leq s_i$ for each $i < n$, then

$$f(r_0, \ldots, r_{n-1}) \leq f(s_0, \ldots, s_{n-1}).$$

This is an immediate consequence of the general distributivity property and the definition of $\leq$. Note: operations of rank 0 are identified with individual constants in A.

An operation f of rank n on the universe of a Boolean algebra A is said to be *completely distributive* if it is completely distributive over addition in each argument in the sense that for each index $i < n$, for each sequence $r_0, \ldots, r_{i-1}, r_{i+1}, \ldots, r_{n-1}$ of elements in A, and for each (finite or infinite) subset X of A, if the supremum $t = \sum X$ exists, then the supremum

$$\sum \{f(r_0, \ldots, r_{i-1}, s, r_{i+1}, \ldots, r_{n-1}) : s \in X\}$$

exists and

$$f(r_0, \ldots, r_{i-1}, t, r_{i+1}, \ldots, r_{n-1})$$
$$= \sum \{f(r_0, \ldots, r_{i-1}, s, r_{i+1}, \ldots, r_{n-1}) : s \in X\}.$$

For example, a binary operation $\circ$ on A is completely distributive if for every subset X of A, if the supremum $t = \sum X$ exists, then the suprema

$$\sum\{r \circ s : s \in X\} \qquad \text{and} \qquad \sum\{s \circ r : s \in X\}$$

both exist, and

$$r \circ t = \sum\{r \circ s : s \in X\} \qquad \text{and} \qquad t \circ r = \sum\{s \circ r : s \in X\}$$

for all r in A. A completely distributive operation on A is called a *complete operator*. Such an operator f always has the following *general complete distributivity property*: for any sequence $X_0, \ldots, X_{n-1}$ of (finite or infinite) subsets of A, if the sum $t_i = \sum X_i$ exists in A for each $i < n$, then the sum on the right side of (1) exists and the equation in (1) holds. The proof involves a straight-forward induction on the rank n of the complete operator. For example, in the case of a binary complete operator $\circ$, we have

$$\begin{aligned} t_0 \circ t_1 &= \sum\{r \circ t_1 : r \in X_0\} \\ &= \sum\{\sum\{r \circ s : s \in X_1\} : r \in X_0\} \\ &= \sum\{r \circ s : r \in X_0 \text{ and } s \in X_1\}. \end{aligned}$$

In [21], a weaker notion of complete distributivity is used in which the set X is always required to be non-empty. We shall refer to operators with this weaker property as *quasi-completely distributive*, or simply *quasi-complete*. Such an operator f always satisfies a restricted version of the general complete distributivity property in which each of the sets in the sequence $X_0, \ldots, X_{n-1}$ is assumed to be non-empty (see Theorem 1.6 in [21]).

An operation f on A of rank n is called *normal* if it always assumes the value 0 when at least one of its arguments is 0, that is to say, if for each $i < n$ and for each sequence

$$r_0, \ldots, r_{i-1}, r_{i+1}, \ldots, r_{n-1} \tag{2}$$

of elements in A,

$$f(r_0, \ldots, r_{i-1}, 0, r_{i+1}, \ldots, r_{n-1}) = 0.$$

Notice that a quasi-complete operator f is complete if and only if it is normal. Indeed, if (2) is a sequence of elements in A, and if a set X is empty, then

$$t = \sum X = \sum \varnothing = 0$$

and

$$\sum\{f(r_0,\ldots,r_{i-1},s,r_{i+1},\ldots,r_{n-1}) : s \in X\} = \sum \varnothing = 0$$

(where $\varnothing$ denotes the empty set), so that

$$f(r_0,\ldots,r_{i-1},t,r_{i+1},\ldots,r_{n-1})$$
$$= \sum\{f(r_0,\ldots,r_{i-1},s,r_{i+1},\ldots,r_{n-1}) : s \in X\}$$

if and only if

$$f(r_0,\ldots,r_{i-1},0,r_{i+1},\ldots,r_{n-1}) = 0.$$

A *Boolean algebra with operators* is an algebra of the form

$$\mathfrak{A} = (A,+,-,f_\xi)_{\xi \in \Xi}, \tag{3}$$

where $(A,+,-)$ is a Boolean algebra—called the *Boolean part* of $\mathfrak{A}$—and f_ξ is an operator on A for each ξ in Ξ. The algebra is called *normal* if each of its operators is normal. It follows from the remark at the end of the preceding paragraph that a Boolean algebra with complete operators is automatically normal. In this monograph, *all Boolean algebras with operators are assumed to be normal*, so we shall not bother to repeat this hypothesis. In other word, when we speak of a Boolean algebra with operators, it will always be understood that the algebra is assumed to be normal. We use upper case German fraktur letters from the beginning of the alphabet to refer to Boolean algebras with operators, and corresponding italic letters to refer to the universes of these algebras. For example, A is assumed to be the universe of a given algebra $\mathfrak{A}$. Such an algebra is said to possess a certain Boolean property if the Boolean part of $\mathfrak{A}$ possesses this property. For instance, $\mathfrak{A}$ is said to be atomic if the Boolean part of $\mathfrak{A}$ is atomic. *An exception is made, however, for the notion of completeness*: $\mathfrak{A}$ is said to be *complete* if the Boolean part of $\mathfrak{A}$ is complete *and* if each of the operators is complete, that is to say, completely distributive.

The *similarity type* of a Boolean algebra with operators is the sequence of ranks of its fundamental operations. If the algebra has the form (3), then the similarity type is $(2,1,n_\xi)_{\xi \in \Xi}$, where n_ξ is the rank of the operator f_ξ. Throughout this paper, *all Boolean algebras with operators will be assumed to have the same arbitrary but fixed similarity type*. In proofs, *we will always deal with one exemplary operator* ∘

that is assumed to be binary. This will free the proofs from excessive notation and hopefully make the main ideas of the proofs clearer to the reader. The passage from the case of a binary operator to the case of an operator of arbitrary rank n is always straightforward, and can safely be left to the reader. Occasionally, the case of a constant—an operator of rank 0—must be treated somewhat differently, and in these cases we will point out the differences.

Another simplification of notation may be helpful as well. In order to distinguish carefully between the operations of different Boolean algebras with operators, say $\mathfrak{A}$ and $\mathfrak{B}$, one should employ different notations to distinguish the fundamental operations of the two algebras, for example, by using superscripts to write

$$\mathfrak{A} = (A\,,\,+^{\mathfrak{A}}\,,\,-^{\mathfrak{A}}\,,\,f_\xi^{\mathfrak{A}})_{\xi \in \Xi} \quad \text{and} \quad \mathfrak{B} = (B\,,\,+^{\mathfrak{B}}\,,\,-^{\mathfrak{B}}\,,\,f_\xi^{\mathfrak{B}})_{\xi \in \Xi}.$$

In practice, the context usually makes clear when the operation symbols in question refer to the operations of $\mathfrak{A}$ and when they refer to the operations of $\mathfrak{B}$; so we shall omit such superscripts when no confusion can arise.

We shall need the following theorem, which characterizes when a bijection between the sets of atoms of two complete and atomic Boolean algebras with operators can be extended to an isomorphism.

Theorem 1.1. *Suppose $\mathfrak{A}$ and $\mathfrak{B}$ are complete and atomic Boolean algebras with operators. A bijection φ from the set of atoms in $\mathfrak{A}$ to the set of atoms in $\mathfrak{B}$ can be extended to an isomorphism from $\mathfrak{A}$ to $\mathfrak{B}$ if and only if the condition*

$$t \le f(r_0, \ldots, r_{n-1}) \quad \textit{if and only if} \quad \varphi(t) \le f(\varphi(r_0), \ldots, \varphi(r_{n-1}))$$

is satisfied for each operator f of rank n and for each sequence of atoms $r_0, \ldots, r_{n-1}, t$ in $\mathfrak{A}$.

Proof. The necessity of the condition is obvious. To establish its sufficiency, suppose that the condition holds. Write X for the set of all atoms in $\mathfrak{A}$, and for each element u in $\mathfrak{A}$, write X_u for the set of atoms in $\mathfrak{A}$ that are below u. In both $\mathfrak{A}$ and $\mathfrak{B}$, each element is the supremum of the set of atoms that it dominates, distinct elements are the suprema of distinct sets of atoms, and every set of atoms has a supremum. One consequence of this observation and the assumption that φ is a bijection between the sets of atoms is that the suprema

$$\textstyle\sum\{\varphi(t) : t \in X_u\} \quad \text{and} \quad \textstyle\sum\{\varphi(t) : t \in X \sim X_u\}$$

are disjoint and sum to the unit in $\mathfrak{B}$, so

$$\sum\{\varphi(t) : t \in X \sim X_u\} = -\left(\sum\{\varphi(t) : t \in X_u\}\right). \tag{1}$$

A second consequence is that the function ψ from $\mathfrak{A}$ to $\mathfrak{B}$ defined by

$$\psi(u) = \sum\{\varphi(t) : t \in X_u\} \tag{2}$$

for each element u in $\mathfrak{A}$ is a bijection from the universe of $\mathfrak{A}$ to the universe of $\mathfrak{B}$.

If Y is an arbitrary set of elements in $\mathfrak{A}$, then the set of atoms below the sum $u = \sum Y$ is the union, over all v in Y, of the set of atoms below v, that is to say,

$$X_u = \bigcup_{v \in Y} X_v, \tag{3}$$

so

$$\psi(u) = \sum\{\varphi(t) : t \in X_u\} = \sum\{\varphi(t) : t \in \bigcup_{v \in Y} X_v\}$$
$$= \sum_{v \in Y} \sum\{\varphi(t) : t \in X_v\} = \sum_{v \in Y} \psi(v)$$

by (2) and (3), the general associative law for Boolean addition, and (2) (with v in place of u). Consequently, ψ preserves arbitrary sums. The set of atoms below a complement $-u$ is the complement in X of the set of atoms below u, that is to say,

$$X_{-u} = X \sim X_u, \tag{4}$$

so

$$\psi(-u) = \sum\{\varphi(t) : t \in X_{-u}\} = \sum\{\varphi(t) : t \in X \sim X_u\}$$
$$= -\left(\sum\{\varphi(t) : t \in X_u\}\right) = -\psi(u),$$

by (2) (with $-u$ in place of u), (4), (1), and (2). Consequently, ψ preserves complements and is therefore a Boolean isomorphism.

It remains to check that ψ preserves each operator. Consider the case of a binary operator $\circ$. Let u and v be elements in $\mathfrak{A}$, and write $w = u \circ v$. It is to be shown that $\psi(w) = \psi(u) \circ \psi(v)$. Because $\mathfrak{A}$ is atomic, we have

$$u = \sum X_u, \qquad v = \sum X_v, \qquad w = \sum X_w. \tag{5}$$

Because $\mathfrak{A}$ is complete, the operator $\circ$ in $\mathfrak{A}$ is complete, by the definition of a complete Boolean algebra with operators, and therefore (using also the first part of (5))

$$u \circ v = \sum\{r \circ s : r \in X_u \text{ and } s \in X_v\}. \tag{6}$$

It follows that

$$X_w = \{t \in X : t \leq r \circ s \text{ for some } r \in X_u \text{ and } s \in X_s\}. \tag{7}$$

Indeed, if t belongs to the left side of (7), then t is an atom below $u \circ v$, by the definitions of w and X_w, and therefore t must be below $r \circ s$ for some r in X_u and s in X_v, by (6) and the fact that t is an atom. Consequently, t must belong to the right side of (7). On the other hand, if t belongs to the right side of (7), then t is below $r \circ s$, for some r in X_u and s in X_v, and therefore

$$t \leq r \circ s \leq u \circ v = w,$$

by the monotony of the operator $\circ$ and the definition of w. Consequently, t belongs to the left side of (7).

From (2), we see that

$$\psi(u) = \sum\{\varphi(r) : r \in X_u\} \quad \text{and} \quad \psi(v) = \sum\{\varphi(s) : s \in X_v\}. \tag{8}$$

Because $\mathfrak{B}$ is complete, the operator $\circ$ in $\mathfrak{B}$ is complete and therefore (using also (8))

$$\psi(u) \circ \psi(v) = \sum\{\varphi(r) \circ \varphi(s) : r \in X_u \text{ and } s \in X_v\}. \tag{9}$$

The assumed condition from the statement of the theorem ensures that

$$t \leq r \circ s \quad \text{if and only if} \quad \varphi(t) \leq \varphi(r) \circ \varphi(s)$$

for all atoms r, s, and t in $\mathfrak{A}$. Since φ is a bijection between the sets of atoms in the two algebras, and since $\mathfrak{B}$ is atomic, the preceding equivalence implies that ψ maps the set of atoms below $r \circ s$ bijectively to the set of atoms below $\varphi(r) \circ \varphi(s)$, so that

$$\varphi(r) \circ \varphi(s) = \sum\{\varphi(t) : t \in X \text{ and } t \leq r \circ s\}$$

for all atoms r and s in $\mathfrak{A}$. Combine this equation with (9) to arrive at

$$\psi(u) \circ \psi(v)$$
$$= \sum\{\varphi(t) : t \in X \text{ and } t \leq r \circ s \text{ for some } r \in X_u \text{ and } s \in X_v\}.$$

In view of (7) and (2), the preceding equation implies that

$$\psi(u) \circ \psi(v) = \sum\{\varphi(t) : t \in X_w\} = \psi(w),$$

as desired.   □

It is worth pointing out that the isomorphism ψ defined in the proof of the preceding theorem is uniquely determined by φ in the following sense: if ρ is any isomorphism from $\mathfrak{A}$ to $\mathfrak{B}$ that agrees with φ on the set of atoms in $\mathfrak{A}$, then $\rho = \psi$.

As was mentioned earlier, the general notion of a Boolean algebra with (not necessarily normal) operators dates back to Jónsson-Tarski [21], and of course some of the most fundamental results about these algebras are due to them. Specific examples of such algebras were, however, known earlier. These include: the algebras of relations of Peirce [28] and Schröder [33], and the closely connected relation algebras of Tarski (see [3], [38], and [22]); the projective algebras of Everett and Ulam [4]; the cylindric algebras of Chin, Tarski, and Thompson (see [17] and [18]); and the algebras implicit in various theories of modal logic.

1.3 Relational Structures and Complex Algebras

A *relational structure* is a structure of the form

$$\mathfrak{U} = (U, R_\xi)_{\xi \in \Xi} \tag{1}$$

in which the *universe* U is an arbitrary set—the empty set is allowed here—and R_ξ is, for each index ξ, a relation of some finite positive rank n on the set U, that is to say, it is a subset of the set of all n-termed sequences of elements from U; the relations R_ξ are referred to as the *relations in* $\mathfrak{U}$, or, sometimes, the *fundamental relations in* $\mathfrak{U}$. We shall use upper case German fraktur letters from the end of the alphabet to refer to relational structures, and corresponding italic letters to refer to the universes of these structures. For example, U is the universe of a given structure $\mathfrak{U}$.

As in the case of algebras, the similarity type of a relational structure is the sequence of ranks of its fundamental relations. Throughout this paper *all relational structures will be assumed to have the same arbitrary but fixed similarity type that will be coordinated with the similarity type of certain Boolean algebras with operators that are constructed from them.* In proofs, *we will always deal with one exemplary fundamental relation R that is assumed to be ternary.* The hope is, again, to make the main ideas of the proofs clearer to the reader by freeing them from excessive notation. The passage from the case of a ternary relation to the case of a relation of arbitrary rank $n \geq 1$ is always straightforward and can safely be left to the reader.

Just as in the case of the fundamental operations of an algebra, one should employ different notations to distinguish the fundamental relations of different relational structures, for example by writing

$$\mathfrak{U} = (U, R_\xi^\mathfrak{U})_{\xi \in \Xi} \qquad \text{and} \qquad \mathfrak{V} = (V, R_\xi^\mathfrak{V})_{\xi \in \Xi}.$$

Since the context usually makes clear when a given relation symbol refers to a relational structure $\mathfrak{U}$ and when it refers to a relational structure $\mathfrak{V}$, we shall usually use the same symbols to refer to the fundamental relations of different structures.

With each atomic Boolean algebra with operators

$$\mathfrak{A} = (A, +, -, f_\xi)_{\xi \in \Xi},$$

we may correlate a relational structure $\mathfrak{U}$ in the following way. The universe U is the set of all atoms in $\mathfrak{A}$, and for each index ξ, if f_ξ is an operator of rank n in $\mathfrak{A}$, then R_ξ is the relation of rank $n+1$ on U that is defined on each sequence of atoms $r_0, \ldots, r_n$ in $\mathfrak{A}$ by

$$R_\xi(r_0, \ldots, r_n) \qquad \text{if and only if} \qquad r_n \leq f_\xi(r_0, \ldots, r_{n-1}). \qquad (2)$$

(The notation $R_\xi(r_0, \ldots, r_n)$ means that the sequence $(r_0, \ldots, r_n)$ belongs to the relation R_ξ.) For example, correlated with a binary operator $\circ$ in $\mathfrak{A}$ is a ternary relation R on U that is defined by

$$R(r, s, t) \qquad \text{if and only if} \qquad t \leq r \circ s$$

for all atoms r, s, and t in $\mathfrak{A}$. The relational structure $\mathfrak{U}$ is often called the *atom structure* of $\mathfrak{A}$.

Conversely, with each relational structure $\mathfrak{U}$ of the form (1), we may correlate an algebra $\mathfrak{A}$ in the following way. The universe A is the set of all subsets of U, the Boolean operations $+$ and $-$ are respectively the set-theoretic operations of union and complement on subsets of U, and for each index ξ, if R_ξ is a relation of rank $n+1$ in $\mathfrak{U}$, then f_ξ is the operation of rank n on subsets of U that is defined on each sequence of subsets $X_0, \ldots, X_{n-1}$ of U by

$$f_\xi(X_0, \ldots, X_{n-1}) = \{t \in U : R_\xi(r_0, \ldots, r_{n-1}, t)$$
$$\text{for some } r_0 \in X_0, \ldots, r_{n-1} \in X_{n-1}\}. \quad (3)$$

If we identify each singleton subset $\{r\}$ of U with the element r itself—as is often done when no confusion can arise—then we may say that f_ξ is the function that is defined on each sequence of elements $r_0, \ldots, r_{n-1}$ in U by

$$f_\xi(r_0, \ldots, r_{n-1}) = \{t \in U : R_\xi(r_0, \ldots, r_{n-1}, t)\},$$

and that is extended to arbitrary sequences of subsets $X_0, \ldots, X_{n-1}$ of U so as to be completely distributive over unions:

$$f_\xi(X_0, \ldots, X_{n-1}) = \bigcup\{f_\xi(r_0, \ldots, r_{n-1}) : r_0 \in X_0, \ldots, r_{n-1} \in X_{n-1}\}.$$

For example, correlated with a ternary relation R in $\mathfrak{U}$ is the binary operation $\circ$ that is defined on elements r and s in U by

$$r \circ s = \{t \in U : R(r, s, t)\},$$

and that is extended to arbitrary subsets X and Y of U by

$$X \circ Y = \bigcup\{r \circ s : r \in X \text{ and } s \in Y\}$$
$$= \{t \in U : R(r, s, t) \text{ for some } r \in X \text{ and } s \in Y\}.$$

The algebra $\mathfrak{A}$ is called the *complex algebra* of the relational structure $\mathfrak{U}$ and is usually denoted by $\mathfrak{Cm}(U)$. (Notice that in order to simplify the notation a bit, we have in this notation identified the structure $\mathfrak{U}$ with its universe U, as is often done in algebra.) The following theorem (which is the first part of Theorem 3.9 in Jónsson-Tarski [21]) summarizes the preceding observations.

Theorem 1.2. *The complex algebra of a relational structure $\mathfrak{U}$ is a complete and atomic Boolean algebra with operators. In particular, the operators in $\mathfrak{Cm}(U)$ are complete.*

For a concrete example of this construction, consider a group

$$\mathfrak{G} = (G, \circ, ^{-1}, \iota).$$

The group may be viewed as a relational structure of type $(3, 2, 1)$ by treating the binary operation $\circ$ as a ternary relation on G that holds for a triple (f, g, h) just in case $h = f \circ g$, by treating the unary operation $^{-1}$ as a binary relation on G that holds for a pair (f, h) just in case $h = f^{-1}$, and by treating the distinguished element ι as a unary relation on G that holds for an element h just in case $h = \iota$. The operators of the complex algebra of $\mathfrak{G}$ are the usual complex operations $\circ$ and $^{-1}$ that are defined on subsets X and Y of the group by

$$X \circ Y = \{h \in G : h = f \circ g \text{ for some } f \in X \text{ and } g \in Y\},$$
$$X^{-1} = \{h \in G : h = f^{-1} \text{ for some } f \in X\},$$

together with the distinguished constant $\{\iota\}$, which is usually identified with the identity element ι of the group.

The notions of the complex algebra of a relational structure and the atom structure of a complete and atomic Boolean algebra with operators seem to have been studied for the first time in Jónsson-Tarski [21]; see also Henkin-Monk-Tarski [18].

1.4 Duality for Algebras

The atom structure of a complete and atomic Boolean algebra with operators $\mathfrak{A}$ is called the (*first*) *dual*, or the *dual relational structure*, of $\mathfrak{A}$, and the complex algebra of a relational structure $\mathfrak{U}$ is called the (*first*) *dual*, or the *dual* (complete and atomic) *Boolean algebra with operators*, or simply the *dual algebra*, of $\mathfrak{U}$. The dual of the dual of the algebra $\mathfrak{A}$ or the relational structure $\mathfrak{U}$ is called the *second dual* of $\mathfrak{A}$ or $\mathfrak{U}$ respectively. Both $\mathfrak{A}$ and $\mathfrak{U}$ are canonically isomorphic to their second duals. The first part of this assertion is contained in the following theorem (which is essentially the second part of Theorem 3.9 in Jónsson-Tarski [21]).

Theorem 1.3. *Every complete and atomic Boolean algebra with operators is isomorphic to its second dual. More precisely, if $\mathfrak{A}$ is a complete and atomic Boolean algebra with operators, and $\mathfrak{U}$ the atom* *structure of* $\mathfrak{A}$, *then* $\mathfrak{A}$ *is isomorphic to the complex algebra of* $\mathfrak{U}$ *via the function that maps each element* r *in* $\mathfrak{A}$ *to the set of atoms below* r.

Proof. Write $\mathfrak{B}$ for the complex algebra of the atom structure $\mathfrak{U}$. The atoms in $\mathfrak{B}$ are the singletons of elements in $\mathfrak{U}$, that is to say, they are the singletons of atoms in $\mathfrak{A}$. Define a function φ from the set of atoms in $\mathfrak{A}$ to the set of atoms in $\mathfrak{B}$ by

$$\varphi(r) = \{r\}$$

for each atom r in $\mathfrak{A}$. Clearly, φ is a bijection between the sets of atoms. To prove that φ satisfies the condition in Theorem 1.1, focus on the case of a binary operator $\circ$ in $\mathfrak{A}$, the corresponding ternary relation R in $\mathfrak{U}$ that is defined in terms of $\circ$, and the corresponding binary operator $\circ$ in $\mathfrak{B}$ that is defined in terms of R. For any atoms r, s, and t in $\mathfrak{A}$, we have

$$
\begin{array}{lll}
t \leq r \circ s & \text{if and only if} & R(r,s,t), \\
& \text{if and only if} & t \in \{r\} \circ \{s\}, \\
& \text{if and only if} & \{t\} \subseteq \{r\} \circ \{s\}, \\
& \text{if and only if} & \varphi(t) \leq \varphi(r) \circ \varphi(s),
\end{array}
$$

by the definition of the relation R in $\mathfrak{U}$ (see (2) in Section 1.3), the definition of the operator $\circ$ in $\mathfrak{B}$ (see (3) in Section 1.3), and the definition of φ. Thus, φ satisfies the condition in Theorem 1.1. Apply the theorem (and its proof) to conclude that φ can be extended to an isomorphism ψ from $\mathfrak{A}$ to $\mathfrak{B}$ that maps each element r in $\mathfrak{A}$ to the set of atoms that are below r. $\square$

Here is the counterpart for relational structures of the preceding theorem. (It is essentially Theorem 2.7.35 in Henkin-Monk-Tarski [18].)

Theorem 1.4. *Every relational structure is isomorphic to its second dual. More precisely, if* $\mathfrak{U}$ *is a relational structure and* $\mathfrak{B}$ *the complex algebra of* $\mathfrak{U}$, *then* $\mathfrak{U}$ *is isomorphic to the atom structure of* $\mathfrak{B}$ *via the function that maps each element* r *in* $\mathfrak{U}$ *to the singleton* $\{r\}$.

Proof. Write $\mathfrak{V}$ for the atom structure of $\mathfrak{B}$. The elements in $\mathfrak{B}$ are the subsets of $\mathfrak{U}$, and the elements in $\mathfrak{V}$ are the atoms in $\mathfrak{B}$, so the elements in $\mathfrak{V}$ must be the singleton subsets of $\mathfrak{U}$. The function ϑ that maps every element r in $\mathfrak{U}$ to the singleton $\{r\}$ is therefore a bijection from the universe of $\mathfrak{U}$ to the universe of $\mathfrak{V}$. To prove that ϑ preserves the fundamental relations of the two structures, focus on the case of a ternary relation R in $\mathfrak{U}$, the corresponding binary operator $\circ$ in $\mathfrak{B}$ that is defined in terms of R, and the corresponding ternary relation R in $\mathfrak{V}$ that is defined in terms of $\circ$. If r, s, and t are elements in $\mathfrak{U}$, then

$$
\begin{aligned}
R(r,s,t) \qquad &\text{if and only if} \qquad t \in \{r\} \circ \{s\}, \\
&\text{if and only if} \qquad \{t\} \subseteq \{r\} \circ \{s\}, \\
&\text{if and only if} \qquad R(\{r\},\{s\},\{t\}), \\
&\text{if and only if} \qquad R(\vartheta(r), \vartheta(s), \vartheta(t)),
\end{aligned}
$$

by the definition of the operator $\circ$ in $\mathfrak{B}$ (see (3) in Section 1.3), the definition of the relation R in $\mathfrak{V}$ (see (2) in Section 1.3), and the definition of the mapping ϑ. Conclusion: ϑ is an isomorphism. $\square$

The preceding theorems imply the following conclusion (see Theorem 3.9 in [21], and Theorem 2.7.35 and Corollary 2.7.36 in [18]).

Corollary 1.5. *Every relational structure $\mathfrak{U}$ is, up to isomorphism, the atom structure of a complete and atomic Boolean algebra with operators, namely the atom structure of the complex algebra of $\mathfrak{U}$. Similarly, every complete and atomic Boolean algebra with operators $\mathfrak{A}$ is, up to isomorphism, the complex algebra of a relational structure, namely the complex algebra of the atom structure of $\mathfrak{A}$.*

This corollary implies that when developing a duality theory for relational structures and complete, atomic Boolean algebras with operators, it suffices to develop a duality theory for relational structures and their complex algebras. We shall make substantial use of this observation in the sequel.

Another consequence of the preceding theorems is that the isomorphism type of a complete and atomic Boolean algebra with operators determines and is determined by the isomorphism type of its atom structure, and conversely (see Corollary 2.7.37(i) in [18]).

Corollary 1.6. *If $\mathfrak{A}$ and $\mathfrak{B}$ are complete and atomic Boolean algebras with operators, then $\mathfrak{A}$ is isomorphic to $\mathfrak{B}$ if and only if the atom structure of $\mathfrak{A}$ is isomorphic to the atom structure of $\mathfrak{B}$.*

Proof. Let $\mathfrak{U}$ and $\mathfrak{V}$ be the atom structures of $\mathfrak{A}$ and $\mathfrak{B}$ respectively. If ϑ is an isomorphism from $\mathfrak{U}$ to $\mathfrak{V}$, then the function φ defined by

$$\varphi(X) = \{\vartheta(r) : r \in X\}$$

for sets X in $\mathfrak{Cm}(U)$ is easily seen to be a Boolean isomorphism from the complex algebra $\mathfrak{Cm}(U)$ to the complex algebra $\mathfrak{Cm}(V)$, and φ preserves all operations on subsets that are definable in terms of the fundamental relations of atom structures, by the isomorphism properties of ϑ. In particular, φ preserves the operators of the complex algebras, because these operators are defined in terms of the fundamental relations and the Boolean notions. Consequently, φ is an isomorphism from $\mathfrak{Cm}(U)$ to $\mathfrak{Cm}(V)$. Since $\mathfrak{A}$ and $\mathfrak{B}$ are isomorphic to $\mathfrak{Cm}(U)$ and $\mathfrak{Cm}(V)$ respectively, by Theorem 1.3, it follows that $\mathfrak{A}$ and $\mathfrak{B}$ must be isomorphic.

On the other hand, if φ is an isomorphism from $\mathfrak{A}$ to $\mathfrak{B}$, then φ maps the set of atoms in $\mathfrak{A}$ bijectively to the set of atoms in $\mathfrak{B}$, and φ preserves all relations on atoms that are definable in terms of the fundamental operations of $\mathfrak{A}$ and $\mathfrak{B}$. In particular, φ preserves the fundamental relations of the atom structures $\mathfrak{U}$ and $\mathfrak{V}$, since these relations are defined in terms of the operations of the algebras. An appropriate restriction of φ is therefore an isomorphism from $\mathfrak{U}$ to $\mathfrak{V}$. $\square$

Here is the analogue for relational structures of the preceding corollary (see Corollary 2.7.37(ii) in [18]).

Corollary 1.7. *If $\mathfrak{U}$ and $\mathfrak{V}$ are relational structures, then $\mathfrak{U}$ is isomorphic to $\mathfrak{V}$ if and only if the complex algebra of $\mathfrak{U}$ is isomorphic to the complex algebra of $\mathfrak{V}$.*

1.5 Duality for Complete Homomorphisms

The duality between complete and atomic Boolean algebras with operators and relational structures carries with it a duality between the structure-preserving mappings. The structure-preserving mappings on the Boolean algebras with operators are homomorphisms that satisfy an additional condition of completeness. If $\mathfrak{A}$ and $\mathfrak{B}$ are Boolean algebras with operators, then a *homomorphism* from $\mathfrak{A}$ to $\mathfrak{B}$ is a function φ from the universe of $\mathfrak{A}$ to the universe of $\mathfrak{B}$ that preserves operations. In other words, if f is an operator of rank n, then

$$\varphi(f(r_0, \ldots, r_{n-1})) = f(\varphi(r_0), \ldots, \varphi(r_{n-1}))$$

for all $r_0, \ldots, r_{n-1}$ in $\mathfrak{A}$, and similarly for the operations of addition and complement. A homomorphism φ is called *complete* if it preserves all sums that happen to exist. This means that if X is a set of elements in $\mathfrak{A}$ with supremum r, then the set of elements

$$\varphi(X) = \{\varphi(s) : s \in X\}$$

in $\mathfrak{B}$ has a supremum, and that supremum is $\varphi(r)$. Symbolically, we may write $\varphi(\sum X) = \sum \varphi(X)$. Notice that a complete homomorphism must automatically also preserve all infima that happen to exist.

The structure-preserving mappings on relational structures are also homomorphisms that satisfy an additional condition of boundedness. Roughly speaking, this condition requires that certain sequences belonging to the fundamental relations of the range structure must be images of sequences belonging to the corresponding fundamental relations of the domain structure.

Definition 1.8. A *homomorphism* from a relational structure $\mathfrak{U}$ to a relational structure $\mathfrak{V}$ is a function ϑ from the universe of $\mathfrak{U}$ to the universe of $\mathfrak{V}$ that preserves the fundamental relations of these structures in the sense that for every fundamental relation R of rank $n+1$, and every sequence $u_0, \ldots, u_n$ of elements in $\mathfrak{U}$, if $R(u_0, \ldots, u_n)$ holds in $\mathfrak{U}$, then $R(\vartheta(u_0), \ldots, \vartheta(u_n))$ holds in $\mathfrak{V}$. A homomorphism ϑ is called *bounded* if for every fundamental relation R of rank $n+1$, and every sequence of elements $r_0, \ldots, r_{n-1}$ in $\mathfrak{V}$ and v in $\mathfrak{U}$, if

$$R(r_0, \ldots, r_{n-1}, \vartheta(v))$$

holds in $\mathfrak{V}$, then there must be elements $u_0, \ldots, u_{n-1}$ in $\mathfrak{U}$ such that $\vartheta(u_i) = r_i$ for $i < n$ and $R(u_0, \ldots, u_{n-1}, v)$ holds in $\mathfrak{U}$. $\square$

As usual, the terms *epimorphism* and *monomorphism* refer to homomorphisms that are onto and one-to-one respectively. Warning: in opposition to the situation for algebras, a bijective homomorphism from one relational structure to another is not necessarily an isomorphism. For a bijective homomorphism ϑ between relational structures to be an isomorphism, the implication in the first part of the definition above must be replaced by an equivalence. In other words, $R(u_0, \ldots, u_n)$ holds in $\mathfrak{U}$ if and only if $R(\vartheta(u_0), \ldots, \vartheta(u_n))$ holds in $\mathfrak{V}$. Notice, however, that a bijective bounded homomorphism is in fact a isomorphism.

The notion of a bounded homomorphism has its roots in modal logic, where the case of relations of rank 2 was studied (under a different name) by Segerberg [34], and in the work of Pierce [29]. The general definition and terminology we have used is from Goldblatt [13].

Every bounded homomorphism from a relational structure $\mathfrak{U}$ to a relational structure $\mathfrak{V}$ induces a complete homomorphism from the complex algebra $\mathfrak{Cm}(V)$ to the complex algebra $\mathfrak{Cm}(U)$ as follows. The theorem without the conclusion of the completeness of the homomorphism φ, and without the assertion that if φ is one-to-one or onto, then ϑ is onto or one-to-one respectively, is essentially due to Goldblatt [13], Theorem 2.3.1; an earlier version that applies to relational structures with a single binary relation is stated in Thomason [39], Theorem 1. (See also Proposition 5.5.1 in Blackburn-de Rijke-Venema [1] and Proposition 5.6 in Venema [40].)

Theorem 1.9. *If $\mathfrak{U}$ and $\mathfrak{V}$ are relational structures, and ϑ a bounded homomorphism from $\mathfrak{U}$ to $\mathfrak{V}$, then the function φ defined on subsets X of $\mathfrak{V}$ by*

$$\varphi(X) = \vartheta^{-1}(X) = \{u \in U : \vartheta(u) \in X\}$$

is a complete homomorphism from $\mathfrak{Cm}(V)$ to $\mathfrak{Cm}(U)$. Moreover, φ is one-to-one if and only if ϑ is onto, and φ is onto if and only if ϑ is one-to-one.

Proof. Straightforward computations show that the function φ defined in the statement of the theorem is a complete Boolean homomorphism. For instance, if X is a subset of $\mathfrak{V}$, then

$$
\begin{array}{lll}
u \in \vartheta^{-1}(\sim X) & \text{if and only if} & \vartheta(u) \in \sim X, \\
& \text{if and only if} & \vartheta(u) \notin X, \\
& \text{if and only if} & u \notin \vartheta^{-1}(X), \\
& \text{if and only if} & u \in \sim \vartheta^{-1}(X),
\end{array}
$$

by the definitions of the complement of a set and the inverse image of a set under a function. Combine these equivalences with the definition of φ to conclude that

$$\varphi(\sim X) = \vartheta^{-1}(\sim X) = \sim \vartheta^{-1}(X) = \sim\varphi(X).$$

Thus, φ preserves complements. The proof that φ preserves arbitrary unions is similar. If $(X_i : i \in I)$ is a system of subsets of $\mathfrak{V}$, then

$$
\begin{array}{lll}
u \in \vartheta^{-1}(\bigcup_i X_i) & \text{if and only if} & \vartheta(u) \in \bigcup_i X_i, \\
& \text{if and only if} & \vartheta(u) \in X_i \text{ for some } i, \\
& \text{if and only if} & u \in \vartheta^{-1}(X_i) \text{ for some } i, \\
& \text{if and only if} & u \in \bigcup_i \vartheta^{-1}(X_i),
\end{array}
$$

by the definitions of the union of a system and the inverse image of a set under a function. Consequently,

$$\varphi(\bigcup_i X_i) = \vartheta^{-1}(\bigcup_i X_i) = \bigcup_i \vartheta^{-1}(X_i) = \bigcup_i \varphi(X_i),$$

so φ preserves arbitrary unions.

To prove that φ preserves the operators of the complex algebras, focus on the case of a ternary relation R and the corresponding binary operator $\circ$. For subsets X and Y of $\mathfrak{V}$, it is to be shown that

$$\varphi(X \circ Y) = \varphi(X) \circ \varphi(Y),$$

or what amounts to the same thing, that

$$\vartheta^{-1}(X \circ Y) = \vartheta^{-1}(X) \circ \vartheta^{-1}(Y). \tag{1}$$

Assume first that an element w belongs to the right side of (1). This means that there must be elements u in $\vartheta^{-1}(X)$ and v in $\vartheta^{-1}(Y)$ such that $R(u, v, w)$ holds in $\mathfrak{U}$, by the definition of the operator $\circ$ in $\mathfrak{Cm}(U)$. The element $\vartheta(u)$ is in X, and $\vartheta(v)$ is in Y, by the definition of the inverse image of a set, and the homomorphism properties of ϑ imply that $R(\vartheta(u), \vartheta(v), \vartheta(w))$ holds in $\mathfrak{V}$. Consequently, $\vartheta(w)$ is in $X \circ Y$, by the definition of the operator $\circ$ in $\mathfrak{Cm}(V)$, and therefore w belongs to the left side of (1).

Assume now that w belongs to the left side of (1). In this case, $\vartheta(w)$ is in the set $X \circ Y$, by the definition of the inverse image of a set, so there must be elements r in X and s in Y such that $R(r, s, \vartheta(w))$ holds in $\mathfrak{V}$, by the definition of the operator $\circ$ in $\mathfrak{Cm}(V)$. The homomorphism ϑ is assumed to be bounded, so there are elements u and v in $\mathfrak{U}$ such that

$$\vartheta(u) = r, \qquad \vartheta(v) = s, \qquad \text{and} \qquad R(u, v, w)$$

in $\mathfrak{U}$, by Definition 1.8. In particular, the elements u and v belong to the sets $\vartheta^{-1}(X)$ and $\vartheta^{-1}(Y)$ respectively (because r and s are in X and Y respectively), and consequently w belongs to the right side of (1), by the definition of the operator $\circ$ in $\mathfrak{Cm}(U)$. This completes the proof that φ is a complete homomorphism.

Turn now to the second assertion of the theorem. Assume first that ϑ is one-to-one, with the aim of showing that φ is onto. Given an arbitrary subset Y of $\mathfrak{U}$, write

$$X = \vartheta(Y) = \{\vartheta(u) : u \in Y\},$$

and observe that

$$\varphi(X) = \vartheta^{-1}(X) = Y,$$

by the definition of φ and the assumption that ϑ is one-to-one. Conclusion: φ maps $\mathfrak{Cm}(V)$ onto $\mathfrak{Cm}(U)$.

Assume next that φ is onto, with the aim of showing that ϑ is one-to-one. To this end, consider distinct elements u and v in $\mathfrak{U}$. Since φ is onto, there must be subsets X and Y of $\mathfrak{V}$ such that

$$\varphi(X) = \vartheta^{-1}(X) = \{u\} \qquad \text{and} \qquad \varphi(Y) = \vartheta^{-1}(Y) = \{v\}.$$

Thus, u is the unique element in $\mathfrak{V}$ that is mapped by ϑ into the set X, and v is the unique element in $\mathfrak{V}$ that is mapped by ϑ into the set Y, by the definition of the inverse image of a set. In particular, since v is different from u, the element u is not mapped by ϑ to any element in Y, while v *is* mapped by ϑ to an element in Y. Consequently, we must have $\vartheta(u) \neq \vartheta(v)$, so ϑ is one-to-one. The observations of this paragraph and the previous one show that the function ϑ is one-to-one if and only if the function φ is onto.

Suppose now that φ is one-to-one, with the aim of showing that ϑ is onto. Given an arbitrary element r in $\mathfrak{V}$, observe that the image

$$\varphi(\{r\}) = \vartheta^{-1}(\{r\})$$

cannot be empty, since φ is a Boolean monomorphism and therefore maps zero, and only zero, to zero. There is consequently an element u in $\mathfrak{U}$ such that $\vartheta(u) = r$, by the definition of the inverse image of a set, so ϑ maps $\mathfrak{U}$ onto $\mathfrak{V}$.

Assume finally that ϑ is onto, with the aim of showing that φ is one-to-one. To this end, consider distinct subsets X and Y of $\mathfrak{V}$. There must be an element belonging to one of these sets but not to the other, say r belongs to X but not to Y. Since ϑ is onto, there is an element u in $\mathfrak{U}$ such that $\vartheta(u) = r$. The element u belongs to the inverse image $\vartheta^{-1}(X)$, because r is in X; but u cannot belong to the inverse image $\vartheta^{-1}(Y)$, because $\vartheta(u) = r$ and r is not in Y. Consequently,

$$\varphi(X) = \vartheta^{-1}(X) \neq \vartheta^{-1}(Y) = \varphi(Y),$$

so φ is one-to-one. The arguments in this paragraph and the previous one show that the function ϑ is onto if and only if the function φ is one-to-one. $\square$

A kind of converse to Theorem 1.9 is also true: every complete homomorphism from a complex algebra $\mathfrak{Cm}(V)$ to a complex algebra $\mathfrak{Cm}(U)$ induces a bounded homomorphism from $\mathfrak{U}$ to $\mathfrak{V}$. A statement of this result, without the epi-mono correspondence, is implicit in Jónsson [19], Theorem 3.1.5, but no proof is given there; an earlier version that applies to relational structures with a single binary relation is stated in Thomason [39], Theorem 1. (See also Proposition 5.7 in Venema [40].)

Theorem 1.10. *If $\mathfrak{U}$ and $\mathfrak{V}$ are relational structures, and φ a complete homomorphism from $\mathfrak{Cm}(V)$ to $\mathfrak{Cm}(U)$, then the function ϑ defined on elements u in $\mathfrak{U}$ by*

$$\vartheta(u) = r \qquad \text{if and only if} \qquad r \in \bigcap\{X \subseteq V : u \in \varphi(X)\},$$

or, equivalently, by

$$\vartheta(u) = r \qquad \text{if and only if} \qquad u \in \varphi(\{r\}),$$

is a bounded homomorphism from $\mathfrak{U}$ to $\mathfrak{V}$. Moreover, ϑ is one-to-one if and only if φ is onto, and ϑ is onto if and only if φ is one-to-one.

Proof. For each element u in $\mathfrak{U}$, write

$$F_u = \{X \subseteq V : u \in \varphi(X)\}, \tag{1}$$

and observe that F_u is a ultrafilter in $\mathfrak{Cm}(V)$. For instance, to see that F_u is closed under intersection, suppose that sets X and Y are both in F_u. In this case, the element u belongs to both of the image sets $\varphi(X)$ and $\varphi(Y)$, by (1), and therefore u is in the intersection of these two image sets. Since

$$\varphi(X \cap Y) = \varphi(X) \cap \varphi(Y),$$

by the homomorphism properties of φ, it follows that u is in $\varphi(X \cap Y)$, and therefore $X \cap Y$ belongs to F_u, by (1). A similar argument shows that if a set X belongs to F_u, and if X is included in a subset Y of V, then Y must also belong to F_u. The empty set $\varnothing$ cannot belong to F_u, because $\varphi(\varnothing) = \varnothing$, by the homomorphism properties of φ, and therefore u cannot be in $\varphi(\varnothing)$. The argument so far shows that F_u is a proper Boolean filter. To see that it is actually an ultrafilter, consider a subset X of V that does not belong to F_u. In this case, the element u is not in $\varphi(X)$, by (1), so u must be in $\sim\varphi(X)$. Since

$$\varphi(\sim X) = \sim\varphi(X),$$

by the homomorphism properties of φ, it follows that u is in $\varphi(\sim X)$, and therefore $\sim X$ belongs to F_u, by (1).

The next step in the proof is to observe that the intersection of the sets in F_u cannot be empty. Indeed,

$$\varphi(\bigcap F_u) = \bigcap\{\varphi(X) : X \in F_u\} = \bigcap\{\varphi(X) : u \in \varphi(X)\},$$

by the assumed completeness of the homomorphism φ, and (1). The element u clearly belongs to the set on the right, so u belongs to the set $\varphi(\bigcap F_u)$. Consequently, this last set is not empty. Since φ is a homomorphism, it follows that the set $\bigcap F_u$ cannot be empty.

Suppose r is an element in the intersection $\bigcap F_u$. Exactly one of the two sets $\{r\}$ and $V \sim \{r\}$ belongs to F_u, by the ultrafilter properties of F_u. The set in F_u cannot be $V \sim \{r\}$, for then r could not belong to the intersection of the sets in F_u. Consequently, the set $\{r\}$ must belong to F_u. Obviously, two distinct singleton sets cannot both belong to F_u, for then the intersection of these two sets would also belong to F_u, contradicting the fact that the empty set is not in F_u.

The observations of the preceding paragraph imply the following two conclusions. First, for every element u in $\mathfrak{U}$ there is a unique element r in $\mathfrak{V}$ that belongs to every set in F_u, so the correspondence ϑ defined in the statement of the theorem is a well-defined function from $\mathfrak{U}$ to $\mathfrak{V}$. Second, the function ϑ can equivalently be defined by writing

$$\vartheta(u) = r \qquad \text{if and only if} \qquad u \in \varphi(\{r\}). \tag{2}$$

Turn now to the task of showing that ϑ is a bounded homomorphism. Focus on the case of a ternary relation R and a corresponding binary operator $\circ$ in the complex algebra. Consider elements u, v, and w in $\mathfrak{U}$, and write

$$\vartheta(u) = r, \qquad \vartheta(v) = s, \qquad \vartheta(w) = t. \tag{3}$$

The equivalent definition of ϑ in (2) implies that

$$u \in \varphi(\{r\}), \qquad v \in \varphi(\{s\}), \qquad w \in \varphi(\{t\}). \tag{4}$$

Assume that $R(u, v, w)$ holds in $\mathfrak{U}$, with the goal of showing that $R(r, s, t)$ holds in $\mathfrak{V}$. The assumption and the definition of the operator $\circ$ in $\mathfrak{Cm}(U)$ imply that

$$w \in \{u\} \circ \{v\}. \tag{5}$$

It follows from (4) that

$$\{u\} \subseteq \varphi(\{r\}) \quad \text{and} \quad \{v\} \subseteq \varphi(\{s\}),$$

and therefore

$$\{u\} \circ \{v\} \subseteq \varphi(\{r\}) \circ \varphi(\{s\}) = \varphi(\{r\} \circ \{s\}),$$

by the monotony of the operator $\circ$ and the homomorphism properties of the function φ. Combine this observation with (5) to conclude that w belongs to $\varphi(\{r\} \circ \{s\})$. In other words, $\{r\} \circ \{s\}$ is one of the sets in the ultrafilter F_w. The element t belongs to every set in F_w, by the definition of ϑ and the final equation in (3). In particular, t belongs to the set $\{r\} \circ \{s\}$, so $R(r, s, t)$ holds in $\mathfrak{V}$, by the definition of the operator $\circ$ in $\mathfrak{Cm}(V)$. Thus, the function ϑ preserves the relation R in the sense of Definition 1.8.

In order to show that ϑ is bounded in the sense of Definition 1.8, consider elements r and s in $\mathfrak{V}$, and w in $\mathfrak{U}$, and write $t = \vartheta(w)$. Assume $R(r, s, t)$ holds in $\mathfrak{V}$, with the goal of finding elements u and v in $\mathfrak{U}$ such that the first two equations in (3) hold and $R(u, v, w)$ is true in $\mathfrak{U}$. The assumption and the definition of the operator $\circ$ in $\mathfrak{Cm}(V)$ imply that t belongs to the set $\{r\} \circ \{s\}$, so that

$$\{t\} \subseteq \{r\} \circ \{s\}.$$

Apply the homomorphism φ to obtain

$$\varphi(\{t\}) \subseteq \varphi(\{r\} \circ \{s\}) = \varphi(\{r\}) \circ \varphi(\{s\}).$$

The element w belongs to the set $\varphi(\{t\})$, by the right side of (4) (which is a consequence of the equivalent definition of ϑ in (2) and the assumption that the equation on the right side of (3) holds). It follows that w belongs to the set $\varphi(\{r\}) \circ \varphi(\{s\})$. This means that there are elements u in $\varphi(\{r\})$ and v in $\varphi(\{s\})$ such that $R(u, v, w)$ holds in $\mathfrak{U}$, by the definition of the operator $\circ$ in $\mathfrak{Cm}(U)$. Consequently, the first two equations in (3) hold, by (2), so ϑ is bounded. The proof of the first assertion of the theorem is thus complete.

Turn now to the second assertion of the theorem. Suppose first that the homomorphism φ is onto. To prove that ϑ is one-to-one, consider distinct elements u and v in $\mathfrak{U}$, with the goal of showing that $\vartheta(u)$

and $\vartheta(v)$ are distinct. Because φ is onto, there must be a subset X of $\mathfrak{V}$ such that $\varphi(X) = \{u\}$. Obviously, u belongs to the image set $\varphi(X)$ while v does not, so the set X belongs to F_u, but not to F_v, by the definitions of F_u and F_v—see (1). Since F_v is an ultrafilter, it follows that $\sim X$ belongs to F_v. The set F_v therefore contains the complement of a set in F_u, so the intersections $\bigcap F_u$ and $\bigcap F_v$ are disjoint. This implies that $\vartheta(u) \neq \vartheta(v)$, by the definition of ϑ.

Assume now that ϑ is one-to-one. To prove that φ is onto, consider an arbitrary subset Y of $\mathfrak{U}$, and put

$$X = \vartheta(Y) = \{\vartheta(u) : u \in Y\}.$$

The goal is to show that $\varphi(X) = Y$. For the proof that Y is included in $\varphi(X)$, consider an arbitrary element u in Y, and write $\vartheta(u) = r$. The element r is in X, by the definition of X, and u is in $\varphi(\{r\})$, by (2), so the singleton $\{r\}$ is included in X, and therefore $\varphi(\{r\})$ is included in $\varphi(X)$, by the homomorphism properties of φ; and consequently, u must belong to $\varphi(X)$. Thus, Y is included in $\varphi(X)$, as claimed. To establish the reverse inclusion, suppose u is an element in $\mathfrak{U}$ that does not belong to Y, with the goal of showing that u cannot belong to $\varphi(X)$. The assumption on u and the fact that ϑ is one-to-one imply that $\vartheta(u) \neq \vartheta(v)$ for every v in Y, so $\vartheta(u)$ is not in the set X, by the definition of X. Write $r = \vartheta(u)$, and observe that u belongs to the set $\varphi(\{r\})$, by (2). Since r is not in X, the sets $\{r\}$ and X are disjoint, and therefore so are the sets $\varphi(\{r\})$ and $\varphi(X)$, by the homomorphism properties of φ. It follows that u cannot belong to the set $\varphi(X)$. Thus, $\varphi(X) = Y$, so φ is onto.

The proof of the dual statement is more straightforward. It is easy to see that each of the following statements is equivalent to its neighbor: (1) φ is one-to-one; (2) $\varphi(\{r\}) \neq \varnothing$ for every atom $\{r\}$ in $\mathfrak{Cm}(V)$; (3) for every element r in $\mathfrak{V}$, there is an element u in $\mathfrak{U}$ that belongs to $\varphi(\{r\})$; (4) for every element r in $\mathfrak{V}$, there is an element u in $\mathfrak{U}$ such that $\vartheta(u) = r$; (5) ϑ is onto. Statements (1) and (2) are equivalent because $\mathfrak{Cm}(V)$ is atomic, and the homomorphism φ is one-to-one just in case its kernel contains only the zero element. The equivalence of (2) and (3) is a consequence of the definition of $\mathfrak{Cm}(V)$, the equivalence of (3) and (4) follows from the equivalent definition of the function ϑ, and the equivalence of (4) and (5) is obvious. Conclusion: φ is one-to-one just in case ϑ is onto. $\square$

The complete homomorphism φ in the statement of Theorem 1.9 is called the *first dual*, or the *dual complete homomorphism*, of the bounded homomorphism ϑ. Notice that the definition of φ may be written in the form

$$u \in \varphi(X) \qquad \text{if and only if} \qquad \vartheta(u) \in X.$$

The bounded homomorphism ϑ in the statement of Theorem 1.10 is called the *first dual*, or the *dual bounded homomorphism*, of the complete homomorphism φ. Observe that for an arbitrary subset X of $\mathfrak{V}$ we have

$$\vartheta(u) \in X \qquad \text{if and only if} \qquad \vartheta(u) = r \quad \text{for some } r \in X,$$
$$\text{if and only if} \qquad u \in \varphi(\{r\}) \quad \text{for some } r \in X,$$
$$\text{if and only if} \qquad u \in \varphi(X),$$

by the definition of the function ϑ and the assumption that the homomorphism φ is complete. Consequently,

$$\vartheta(u) \in X \qquad \text{if and only if} \qquad u \in \varphi(X),$$

and therefore
$$\varphi(X) = \vartheta^{-1}(X).$$

This equation is of course the definition of φ in terms of ϑ in Theorem 1.9. The point is that the equation is also valid in the context of Theorem 1.10, where φ is given and ϑ is being defined.

If φ is a complete homomorphism from $\mathfrak{Cm}(V)$ to $\mathfrak{Cm}(U)$, and ϑ the dual bounded homomorphism from $\mathfrak{U}$ to $\mathfrak{V}$, then the dual of ϑ is called the *second dual* of φ. This second dual is, by Theorem 1.9, the complete homomorphism φ^* from $\mathfrak{Cm}(V)$ to $\mathfrak{Cm}(U)$ that is defined on subsets X of $\mathfrak{V}$ by

$$\varphi^*(X) = \vartheta^{-1}(X).$$

Since the set $\vartheta^{-1}(X)$ coincides with $\varphi(X)$, by the observations of the preceding paragraph, it follows that

$$\varphi^*(X) = \varphi(X)$$

for all subsets X of $\mathfrak{V}$. In other words, φ is its own second dual.

Similarly, if ϑ is a bounded homomorphism from $\mathfrak{U}$ to $\mathfrak{V}$, and φ the dual complete homomorphism from $\mathfrak{Cm}(V)$ to $\mathfrak{Cm}(U)$, then the dual of φ is called the *second dual* of ϑ. This second dual is, by Theorem 1.10, the bounded homomorphism ϑ^* from $\mathfrak{U}$ to $\mathfrak{V}$ that is defined on all elements u in $\mathfrak{U}$ by

$$\vartheta^*(u) = r \qquad \text{if and only if} \qquad u \in \varphi(\{r\}).$$

In view of the definition of φ in terms of ϑ in Theorem 1.9, we have

$$\varphi(\{r\}) = \vartheta^{-1}(\{r\})$$

for every element r in $\mathfrak{V}$, so that each of the statements

$$u \in \varphi(\{r\}), \qquad u \in \vartheta^{-1}(\{r\}), \qquad \vartheta(u) \in \{r\}, \qquad \vartheta(u) = r$$

is equivalent to its neighbor. Combine these remarks to conclude that

$$\vartheta^*(u) = \vartheta(u)$$

for every element u in $\mathfrak{U}$. In other words, ϑ is its own second dual.

Suppose next that $\mathfrak{U}$, $\mathfrak{V}$, and $\mathfrak{W}$ are relational structures, and ϑ a bounded homomorphism from $\mathfrak{U}$ to $\mathfrak{V}$, and δ a bounded homomorphism from $\mathfrak{V}$ to $\mathfrak{W}$. The composition $\delta \circ \vartheta$ is easily seen to be a bounded homomorphism from $\mathfrak{U}$ to $\mathfrak{W}$ (see the diagram below). The dual of ϑ is the complete homomorphism φ from $\mathfrak{Cm}(V)$ to $\mathfrak{Cm}(U)$ that is defined by

$$\varphi(X) = \vartheta^{-1}(X)$$

for elements X in $\mathfrak{Cm}(V)$, and the dual of δ is the complete homomorphism ψ from $\mathfrak{Cm}(W)$ to $\mathfrak{Cm}(V)$ that is defined by

$$\psi(Y) = \delta^{-1}(Y)$$

for elements Y in $\mathfrak{Cm}(W)$. The composition $\varphi \circ \psi$ is a complete homomorphism from $\mathfrak{Cm}(W)$ to $\mathfrak{Cm}(U)$, and

$$(\varphi \circ \psi)(Y) = \varphi(\psi(Y)) = \vartheta^{-1}(\delta^{-1}(Y)) = (\delta \circ \vartheta)^{-1}(Y)$$

for all Y in $\mathfrak{Cm}(W)$. Consequently, the composition $\varphi \circ \psi$ is the dual of the composition $\delta \circ \vartheta$, by the definition of that dual (see the diagram below).

$$\mathfrak{U} \xrightarrow{\ \vartheta\ } \mathfrak{V} \xrightarrow{\ \delta\ } \mathfrak{W}$$

$$\mathfrak{Cm}(W) \xrightarrow{\ \psi\ } \mathfrak{Cm}(V) \xrightarrow{\ \varphi\ } \mathfrak{Cm}(U)$$

The results of this section may be summarized as follows.

Theorem 1.11. *Let* $\mathfrak{U}$, $\mathfrak{V}$, *and* $\mathfrak{W}$ *be relational structures. There is a bijective correspondence between the set of bound homomorphisms* ϑ *from* $\mathfrak{U}$ *to* $\mathfrak{V}$ *and the set of complete homomorphisms* φ *from* $\mathfrak{Cm}(V)$ *to* $\mathfrak{Cm}(U)$ *such that the equivalence*

$$u \in \varphi(X) \qquad \text{if and only if} \qquad \vartheta(u) \in X$$

holds for all subsets X *of* $\mathfrak{V}$ *and all elements* u *in* $\mathfrak{U}$. *Each of the mappings* ϑ *and* φ *is its own second dual. The bounded homomorphism* ϑ *is one-to-one if and only if its dual complete homomorphism* φ *is onto, and* ϑ *is onto if and only if* φ *is one-to-one. If* δ *is a bounded homomorphism from* $\mathfrak{V}$ *to* $\mathfrak{W}$, *with dual* ψ, *then the dual of the composition* $\delta \circ \vartheta$ *is the composition* $\varphi \circ \psi$.

Part of the contents of the preceding theorem may be expressed by saying that the correspondence taking each relational structure to its complex algebra, and each bounded homomorphism to its dual complete homomorphism, is a contravariant functor from the category of all relational structures with bounded homomorphisms as morphisms to the category of all complex algebras with complete homomorphisms as morphisms; and the correspondence taking each complex algebra to its underlying relational structure, and each complete homomorphism to its dual bounded homomorphism, is a contravariant function from the category of all complex algebras with complete homomorphisms as morphisms to the category of all relational structures with bounded homomorphisms as morphisms; and these two contravariant functors are inverses of one another. Consequently, the two categories are said to be *dually equivalent*. This statement occurs without proof as Theorem 3.1.5 in the survey article Jónsson [19]; see also Theorem 5.8 in the survey article Venema [40]. An earlier version that applies to the category of relational structures with a single binary relation (the category of frames with bounded homomorphisms) and the category of complete and atomic Boolean algebras with a single complete unary operator (the category of complete and atomic modal algebras with complete homomorphisms) is formulated in Theorem 1 of Thomason [39].

1.6 Duality for Complete Ideals

The epi-mono duality for structure-preserving mappings, developed in the preceding section, implies a useful sub-quotient duality for the structures themselves. The "substructures" of a Boolean algebra with operators $\mathfrak{A}$ are just subalgebras in the standard general algebraic sense of the word. A substructure of an arbitrary relational structure is usually defined to be a subset of the universe of the structure, together with the appropriate restrictions of the fundamental relations of the structure. This notion is not quite strong enough for the purposes of an algebraic duality theory. An additional condition is needed that requires certain elements to belong to the universe of the substructure.

Definition 1.12. A *substructure* of a relational structure $\mathfrak{U}$ is a relational structure $\mathfrak{V}$ such that the universe of $\mathfrak{V}$ is a subset of the universe of $\mathfrak{U}$, and the fundamental relations of $\mathfrak{V}$ are the restrictions of the corresponding fundamental relations of $\mathfrak{U}$ in the sense that for every fundamental relation R of rank $n + 1$, and for every sequence of elements $r_0, \ldots, r_n$ in $\mathfrak{V}$, the relationship $R(r_0, \ldots, r_n)$ holds in $\mathfrak{V}$ if and only if it holds in $\mathfrak{U}$. A substructure $\mathfrak{V}$ is said to be an *inner substructure* of $\mathfrak{U}$ if for every fundamental relation R of rank $n + 1 > 1$, and for every sequence of elements $r_0, \ldots, r_{n-1}$ in $\mathfrak{U}$ and t in $\mathfrak{V}$, if $R(r_0, \ldots, r_{n-1}, t)$ holds in $\mathfrak{U}$, then the elements $r_0, \ldots, r_{n-1}$ are all in $\mathfrak{V}$. $\square$

The universe of an inner substructure of $\mathfrak{U}$ is called an *inner subuniverse* of $\mathfrak{U}$. Thus, a subset V of $\mathfrak{U}$ is an inner subuniverse of $\mathfrak{U}$ if and only if for every sequence $r_0, \ldots, r_{n-1}$ of elements in $\mathfrak{U}$, and every element t in V, if $R(r_0, \ldots, r_{n-1}, t)$ holds in $\mathfrak{U}$, then the elements $r_0, \ldots, r_{n-1}$ are all in V. If V is an inner subuniverse of $\mathfrak{U}$, then the corresponding inner substructure of $\mathfrak{U}$ is just the relational structure $\mathfrak{V}$ whose universe is V and whose fundamental relations are the restriction to V of the fundamental relations in $\mathfrak{U}$. The notion of an inner substructure is developed in Goldblatt [13]; the notion itself (in a more general framework) and the terminology date back to Feferman [5].

Definition 1.12 is so constructed that inner substructures are just images of bounded monomorphisms.

Lemma 1.13. *Let $\mathfrak{U}$ and $\mathfrak{V}$ be relational structures. If ϑ is a bounded homomorphism from $\mathfrak{V}$ into $\mathfrak{U}$, then the restriction of $\mathfrak{U}$ to the range of ϑ is an inner substructure of $\mathfrak{U}$. In the reverse direction, if ϑ is an isomorphism from $\mathfrak{V}$ to an inner substructure of $\mathfrak{U}$, then ϑ is a bounded monomorphism from $\mathfrak{V}$ into $\mathfrak{U}$.*

Proof. Focus on the case of a ternary relation R. Suppose first that ϑ is a bounded homomorphism from $\mathfrak{V}$ into $\mathfrak{U}$, and let $\mathfrak{W}$ be the restriction of $\mathfrak{U}$ to the range of ϑ. Clearly, $\mathfrak{W}$ is a substructure of $\mathfrak{U}$, by the definition of a restriction. To prove that $\mathfrak{W}$ is actually an inner substructure, consider elements r and s in $\mathfrak{U}$, and t in $\mathfrak{W}$. Since ϑ maps $\mathfrak{V}$ onto $\mathfrak{W}$, there must be an element w in $\mathfrak{V}$ such that $\vartheta(w) = t$. If $R(r, s, t)$ holds in $\mathfrak{U}$, then obviously

$$R(r, s, \vartheta(w)) \tag{1}$$

holds in $\mathfrak{U}$, so there are elements u and v in $\mathfrak{V}$ such that

$$\vartheta(u) = r, \qquad \vartheta(v) = s, \qquad \text{and} \qquad R(u, v, w),$$

by the assumption that ϑ is bounded. It follows that r and s belong to the image of $\mathfrak{V}$ under ϑ, which is just $\mathfrak{W}$, so $\mathfrak{W}$ is an inner substructure of $\mathfrak{U}$.

Assume now that ϑ an isomorphism from $\mathfrak{V}$ to an inner substructure $\mathfrak{W}$ of $\mathfrak{U}$. To prove that ϑ is bounded, consider elements r and s in $\mathfrak{U}$, and w in $\mathfrak{V}$, such that (1) holds in $\mathfrak{U}$. The element $\vartheta(w)$ belongs to $\mathfrak{W}$, and $\mathfrak{W}$ is an inner substructure of $\mathfrak{U}$, so the elements r and s must belong to $\mathfrak{W}$, by Definition 1.12. Since ϑ maps $\mathfrak{V}$ onto $\mathfrak{W}$, there must be elements u and v in $\mathfrak{V}$ such that $\vartheta(u) = r$ and $\vartheta(v) = s$. Obviously, $R(\vartheta(u), \vartheta(v), \vartheta(w))$ holds in $\mathfrak{U}$, and therefore also in $\mathfrak{W}$, by (1), so $R(u, v, w)$ holds in $\mathfrak{V}$, by the isomorphism properties of ϑ. It follows that ϑ is bounded. $\square$

By taking for ϑ the identity function on the universe of $\mathfrak{V}$ in the preceding lemma, we arrive at the following characterization of inner substructures.

Corollary 1.14. *A relational structure $\mathfrak{V}$ is an inner substructure of a relational structure $\mathfrak{U}$ if and only if the identity function on the universe of $\mathfrak{V}$ is a bounded monomorphism from $\mathfrak{V}$ into $\mathfrak{U}$.*

The preceding characterization occurs as Lemma 3.2.2 in Goldblatt [13]. Note, however, that Goldblatt uses the characterizing condition of the corollary as the definition of an inner substructure, and therefore his characterization says that $\mathfrak{V}$ is an inner substructure if and only if it satisfies the conditions of Definition 1.12.

It is occasionally more convenient to speak of the inner subuniverses of a relational structure instead of the inner substructures themselves. For example, the inner substructures of a relational structure $\mathfrak{U}$ form a natural lattice under the partial ordering relation of being an inner substructure. If one considers the inner subuniverses instead of the inner substructures themselves, then the partial ordering relation of the lattice is just the set-theoretic relation of inclusion. In more detail, if $\mathfrak{V}$ and $\mathfrak{W}$ are inner substructures of a relational structure $\mathfrak{U}$, then $\mathfrak{W}$ is an inner substructure of $\mathfrak{V}$ if and only if universe of $\mathfrak{W}$ is included in the universe of $\mathfrak{V}$. It is a straightforward matter to pass between inner substructures and inner subuniverses.

The complete homomorphic images of a complex algebra $\mathfrak{Cm}(U)$ do not form a natural lattice; but every such homomorphic image is isomorphic to the quotient of $\mathfrak{Cm}(U)$ modulo a complete ideal in $\mathfrak{Cm}(U)$, and the complete ideals in $\mathfrak{Cm}(U)$ do form a natural lattice under the relation of inclusion. The duality between complete epimorphisms and bounded monomorphism implies a corresponding duality between complete ideals and inner subuniverses, and this duality induces a dual isomorphism from the lattice of complete ideals in $\mathfrak{Cm}(U)$ to the lattice of inner subuniverses of $\mathfrak{U}$.

Definition 1.15. An *ideal* in a Boolean algebra with operators $\mathfrak{A}$ is a subset M of $\mathfrak{A}$ with the following properties.

  (i) 0 is in M.
 (ii) If r and s are in M, then $r + s$ is in M.
(iii) If r is in M and s in $\mathfrak{A}$, then $r \cdot s$ is in M.
(iv) If f is an operator of rank n in $\mathfrak{A}$, and if $r_0, \ldots, r_{n-1}$ is any
      sequence of elements in $\mathfrak{A}$ such that r_i is in M for some $i < n$,
      then $f(r_0, \ldots, r_{n-1})$ is in M.

An ideal M is called *complete* if whenever $(r_i : i \in I)$ is a system of elements in M with a supremum r in $\mathfrak{A}$, then the supremum r belongs to M. $\quad\square$

The definition of an ideal in a Boolean algebra with operators was originally discovered by Sain [31], who treated the special case when the operators are all unary. Proposition 7.4 in that paper proves that every ideal in a Boolean algebra with unary operators $\mathfrak{A}$ determines a congruence relation of which it is the kernel—the congruence class of 0—and that the kernel of every congruence relation on $\mathfrak{A}$ is an ideal. The general definition was independently discovered some years later by the author and published in [9], with an acknowledgement of Sain's priority.

It is not difficult, on the basis of Definition 1.15, to develop the entire theory of ideals and complete ideals for Boolean algebras with operators. We shall have a bit more to say about this in the next section, but we note here that the set of ideals in $\mathfrak{A}$ is a complete, compactly generated, distributive lattice. The meet of a system of ideals is the intersection of the system, and the join of the system is the intersection of all ideals that include each ideal in the system. It is also not difficult to show that the join and meet of two ideals in this lattice coincide with their complex Boolean sum and product respectively. The set of all complete ideals in $\mathfrak{A}$ is also a lattice. The meet of a system of complete ideals is the intersection of the system, and the join of the system is the intersection of all those complete ideals that include each ideal in the system. As usual, the ideal *generated* by a set W in $\mathfrak{A}$ is defined to be the intersection of the set of all ideals in $\mathfrak{A}$ that include W, and analogously for the complete ideal generated by W.

Consider now a relational structure $\mathfrak{U}$, which is assumed to be fixed throughout the subsequent discussion. Every complete ideal M in the complex algebra $\mathfrak{Cm}(U)$ is *principal* in the sense that there is a maximum element W in M, and M is the set of all elements X in $\mathfrak{Cm}(U)$ that are below (that is to say, included in) W. In fact, if W is the union of the sets in M, then W is an element in $\mathfrak{Cm}(U)$ and therefore W belongs to M, by the assumption that M is complete; and M is the set of all subsets of W, by condition (iii) in the definition of an ideal. The set W is called the *principal generator* of M.

Lemma 1.16. *If M is a complete ideal in $\mathfrak{Cm}(U)$, and if W is the principal generator of M, then the set $V = {\sim}W$ is an inner subuniverse of $\mathfrak{U}$.*

Proof. Focus on the case of a ternary relation R in $\mathfrak{U}$ and the corresponding binary operator $\circ$ in $\mathfrak{Cm}(U)$. Consider elements r and s in $\mathfrak{U}$, and t in V, and assume

$$R(r, s, t) \tag{1}$$

holds in $\mathfrak{U}$. In this case, neither of the sets $\{r\}$ and $\{s\}$ can belong to M. Indeed, if one of the sets, say $\{r\}$, were in M, then the product $\{r\} \circ \{s\}$ would belong to M, by condition (iv) in Definition 1.15. This product is defined to be the set

$$\{r\} \circ \{s\} = \{p : R(r, s, p)\}$$

in $\mathfrak{Cm}(U)$, so t would belong to this product, by (1), and therefore t would belong to W, by the definition of W as the union of the sets in M. However, t cannot belong to W because it is assumed to belong to V, which is the complement of W.

Since neither of the sets $\{r\}$ and $\{s\}$ is in M, neither of the elements r and s can belong to the principal generator W, by the downward closure of M. It follows that both elements must belong to the complement of W, which is V. Thus, V is an inner subuniverse of $\mathfrak{U}$, as claimed. $\square$

There is a natural converse of Lemma 1.16 that is also true.

Lemma 1.17. *If V is an inner subuniverse of $\mathfrak{U}$, then the set $W = {\sim}V$ generates a complete ideal in $\mathfrak{Cm}(U)$ that consists of all subsets of W.*

Proof. It is a straightforward matter to check that the set M of all subsets of W is a complete *Boolean* ideal in $\mathfrak{Cm}(U)$. To verify condition (iv) of Definition 1.15, focus on the case of a binary operator $\circ$ corresponding to a ternary relation R in $\mathfrak{U}$. Suppose X is an element in M, and Y an element in $\mathfrak{Cm}(U)$, with the goal of showing that the product $X \circ Y$ belongs to M. The definition of the operator $\circ$ in $\mathfrak{Cm}(U)$ implies that

$$X \circ Y = \bigcup\{\{r\} \circ \{s\} : r \in X \text{ and } s \in Y\}.$$

The set M is closed under arbitrary unions (because M is complete), so it suffices to show that the product

$$\{r\} \circ \{s\} = \{t : R(r,s,t) \text{ for some } r \in X \text{ and } s \in Y\} \qquad (1)$$

belongs to M for every r in X and s in Y.

The assumption that X belongs to M implies that X is a subset of W, by the definition of M, and therefore X is disjoint from V, which is the complement of W. Consequently, no element t in (1) can belong to V; for if it did, then r and s would have to be in V, by the assumptions that $R(r,s,t)$ holds and that V is an inner subuniverse of $\mathfrak{U}$, and this would contradict the disjointness of X and V. It follows from this observation that the product $\{r\} \circ \{s\}$ must be a subset of ${\sim}V$, and therefore an element in M. Conclusion: $X \circ Y$ belongs to M. A similar argument shows that $Y \circ X$ is also in M. Thus, M satisfies the conditions in Definition 1.15 for being a complete ideal. $\square$

If M is a complete ideal in $\mathfrak{Cm}(U)$, say with principal generator W, then the inner subuniverse of $\mathfrak{U}$ that is the complement of W is called the (*first*) *dual*, or the *dual inner subuniverse*, of M. Inversely, if V is an inner subuniverse of $\mathfrak{U}$, then the principal ideal in $\mathfrak{Cm}(U)$ that is generated by the complement of V is called the (*first*) *dual*, or the *dual complete ideal*, of V.

Start with a complete ideal M in $\mathfrak{Cm}(U)$, say with principal generator W. Thus, M consists of all subsets of W. The dual of M is, by definition, the set $V = {\sim}W$. The dual of V is defined to be the complete ideal of all subsets of ${\sim}V$. Since ${\sim}V = W$, it follows that the dual of V must coincide with M. This can be expressed by saying that the second dual of every complete ideal in $\mathfrak{Cm}(U)$ is itself.

Now start with an inner subuniverse V of $\mathfrak{U}$. The dual of V is the principal ideal M generated by the set $W = {\sim}V$, that is to say, M is the complete ideal of all subsets of W. The dual of M is the inner subuniverse of $\mathfrak{U}$ that is the complement of the principal generator W of M. Since ${\sim}W = V$, it follows that the dual of M coincides with V itself. Thus, the second dual of every inner subuniverse of $\mathfrak{U}$ is itself.

If M and N are complete ideals in $\mathfrak{Cm}(U)$, and V and W their respective dual inner subuniverses, then

$$M \subseteq N \qquad \text{if and only if} \qquad {\sim}V \subseteq {\sim}W,$$
$$\text{if and only if} \qquad W \subseteq V,$$

because ${\sim}V$ and ${\sim}W$ are the principal generators of M and N respectively. It follows that the function φ mapping each complete ideal to its dual inner subuniverse is order reversing in the sense that

$$M \subseteq N \qquad \text{if and only if} \qquad \varphi(N) \subseteq \varphi(M),$$

and is therefore a *dual lattice isomorphism*, that is to say, it is a bijection between the lattices that reverses the lattice partial ordering relation. The following theorem has been proved.

Theorem 1.18. *For a relational structure $\mathfrak{U}$, the dual of every complete ideal in $\mathfrak{Cm}(U)$ is an inner subuniverse of $\mathfrak{U}$, and the dual of every inner subuniverse of $\mathfrak{U}$ is a complete ideal in $\mathfrak{Cm}(U)$. The second dual of every complete ideal and of every inner subuniverse is itself. The function mapping each complete ideal to its dual inner subuniverse is a dual lattice isomorphism from the lattice of complete ideals in $\mathfrak{Cm}(U)$ to the lattice of inner subuniverses of $\mathfrak{U}$.*

1.7 Duality for Complete Quotients

The duality between complete ideals and inner subuniverses implies a duality between the corresponding structures. To formulate this duality, it is necessary to have the notion of a quotient algebra. The quotient of a Boolean algebra with operators $\mathfrak{A}$ modulo an ideal M is defined in the standard way as the algebra of cosets of M. The elements of the quotient are the *cosets*

$$r/M = \{r \ominus s : s \in M\},$$

where $\ominus$ is the Boolean operation of symmetric difference. If f is an operator of rank n in $\mathfrak{A}$, then the corresponding operator in the quotient is defined by

$$f(r_0/M, \ldots, r_{n-1}/M) = f(r_0, \ldots, r_{n-1})/M$$

for elements $r_0, \ldots, r_{n-1}$ in $\mathfrak{A}$, and similarly for the operations of addition and complement. We write $\mathfrak{A}/M$ to denote the quotient algebra.

The standard theorems regarding quotients—in particular, the Homomorphism Theorem and the First Isomorphism Theorem—can be proved in the standard way. In more detail, every ideal M in $\mathfrak{A}$ is the kernel of some epimorphism on $\mathfrak{A}$ in the sense that M is the set of elements mapped to 0 by the epimorphism. In fact, if φ is the *quotient mapping* of $\mathfrak{A}$ onto $\mathfrak{A}/M$—that is to say, if φ is the function that maps each element r in $\mathfrak{A}$ to the coset r/M—then φ is an epimorphism with kernel M. Moreover, if ψ is *any* epimorphism from $\mathfrak{A}$ to a Boolean algebra with operators $\mathfrak{B}$, then the kernel of ψ is an ideal M in $\mathfrak{A}$, and the quotient $\mathfrak{A}/M$ is isomorphic to $\mathfrak{B}$ via the function that maps each coset r/M to the element $\psi(r)$ in $\mathfrak{B}$.

A similar development holds for complete ideals. In particular, the kernel of every complete epimorphism on $\mathfrak{A}$ is a complete ideal in $\mathfrak{A}$, and every complete ideal M in $\mathfrak{A}$ is the kernel of a complete epimorphism on $\mathfrak{A}$, namely the quotient mapping from $\mathfrak{A}$ to $\mathfrak{A}/M$.

Theorem 1.19. *Suppose $\mathfrak{V}$ is an inner substructure of a relational structure $\mathfrak{U}$, and M is the dual complete ideal of the universe of $\mathfrak{V}$. The dual algebra of $\mathfrak{V}$ is isomorphic to the quotient of the dual algebra of $\mathfrak{U}$ modulo the ideal M, that is to say, the complex algebra $\mathfrak{Cm}(V)$ is isomorphic to quotient $\mathfrak{Cm}(U)/M$. Inversely, the dual relational structure of the quotient algebra $\mathfrak{Cm}(U)/M$ is isomorphic to $\mathfrak{V}$.*

Proof. The identity function ϑ on the inner substructure $\mathfrak{V}$ is a bounded monomorphism from $\mathfrak{V}$ into $\mathfrak{U}$, by Corollary 1.14. The dual of ϑ is the complete epimorphism φ from $\mathfrak{Cm}(U)$ to $\mathfrak{Cm}(V)$ that is defined by

$$\varphi(X) = \vartheta^{-1}(X) = X \cap V$$

for elements X in $\mathfrak{Cm}(U)$, by Theorem 1.9 and the definition of ϑ as the identity function on $\mathfrak{V}$. The kernel of φ is the set of elements in $\mathfrak{Cm}(U)$ that are mapped to the empty set. Since

$$\varphi(X) = \varnothing \qquad \text{if and only if} \qquad X \cap V = \varnothing$$
$$\text{if and only if} \qquad X \subseteq {\sim}V,$$

it follows that the kernel of φ is just the dual complete ideal of V, which is M. Consequently, the quotient $\mathfrak{Cm}(U)/M$ is isomorphic to $\mathfrak{Cm}(V)$ via the function that maps each coset X/M to the intersection $X \cap V$, by the First Isomorphism Theorem for Boolean algebras with operators.

Invoke Corollary 1.6 to conclude from the preceding paragraph that the dual relational structure of the quotient $\mathfrak{Cm}(U)/M$ is isomorphic to the dual relational structure of $\mathfrak{Cm}(V)$. Since the dual relational structure of $\mathfrak{Cm}(V)$ is the atom structure of $\mathfrak{Cm}(V)$, and the latter is isomorphic to $\mathfrak{V}$, by Theorem 1.4, it follows that the dual relational structure of $\mathfrak{Cm}(U)/M$ is isomorphic to $\mathfrak{V}$. $\square$

For some remarks concerning the relationship of Theorem 1.19 to the duality between inner substructures and complete homomorphic images that is known from the literature, see the end of Section 1.9.

1.8 Duality for Complete Subuniverses

The second half of the sub-quotient duality concerns quotients of relational structures and complete subalgebras of the dual complex algebras. There is a natural dual isomorphism between the lattice of bounded congruence relations on a relational structure $\mathfrak{U}$ and the lattice of complete subuniverses of $\mathfrak{Cm}(U)$. The notions of a bounded congruence relation and the corresponding quotient relational structure are similar in spirit to the notions of a congruence relation on an algebra and the corresponding quotient algebra. The precise definitions are facilitated by adopting some of the notation used for congruences on algebras. For example, if Θ is an equivalence relation, then we write

$$u \equiv v \mod \Theta$$

to express that the pair (u, v) is in the relation Θ.

Definition 1.20. A *congruence relation*—or, more simply, a *congruence*—on a relational structure $\mathfrak{U}$ is just an equivalence relation on the universe of $\mathfrak{U}$. A congruence Θ on $\mathfrak{U}$ is said to be *bounded* if for every fundamental relation R in $\mathfrak{U}$ of rank $n+1$, and every sequence of elements $u_0, \ldots, u_{n-1}, w, \bar{w}$ in $\mathfrak{U}$, if $R(u_0, \ldots, u_{n-1}, w)$ holds in $\mathfrak{U}$, and if $w \equiv \bar{w} \mod \Theta$, then there are elements $\bar{u}_0, \ldots, \bar{u}_{n-1}$ in $\mathfrak{U}$ such that $u_i \equiv \bar{u}_i \mod \Theta$ for $i < n$, and $R(\bar{u}_0, \ldots, \bar{u}_{n-1}, \bar{w})$ holds in $\mathfrak{U}$. $\quad\square$

A *subalgebra* of a Boolean algebra with operators $\mathfrak{A}$ is an algebra $\mathfrak{B}$ whose universe is a subset of the universe of $\mathfrak{A}$ and whose operations are just the restrictions of the operations of $\mathfrak{A}$ to the universe of $\mathfrak{B}$. A *subuniverse* of $\mathfrak{A}$ is defined to be the universe of a subalgebra of $\mathfrak{A}$. A subset B of $\mathfrak{A}$ is a subuniverse just in case B is non-empty and is closed under the operations of $\mathfrak{A}$. This means that if f is an operator of rank n in $\mathfrak{A}$, then $f(r_0, \ldots, r_{n-1})$ belongs to B whenever $r_0, \ldots, r_{n-1}$ are elements in B, and similarly for the operations of addition and complement. A subalgebra $\mathfrak{B}$ of a complete Boolean algebra with operators $\mathfrak{A}$ is called a *complete subalgebra* if the supremum (in $\mathfrak{A}$) of every subset of $\mathfrak{B}$ belongs to $\mathfrak{B}$. A *complete subuniverse* of $\mathfrak{A}$ is defined to be the universe of a complete subalgebra of $\mathfrak{A}$. In other words, a complete subuniverse of $\mathfrak{A}$ is a non-empty subset of $\mathfrak{A}$ that is closed under the operations of $\mathfrak{A}$ and that contains the supremum of each of its subsets.

The main task of this section is to establish a connection between bounded congruences and complete subuniverses. Consider a congruence relation Θ (not necessarily bounded) on a relational structure $\mathfrak{U}$. A subset X of $\mathfrak{U}$ is said to be *compatible* with Θ if X is a union of equivalence classes of Θ, or what amounts to the same thing, if the presence of an element u in X implies that the entire equivalence class u/Θ is included in X. The empty set and the universal set U are clearly compatible with Θ, and it is easy to check that if a set X is compatible with Θ, then so is the complement $\sim X$, and if sets X_i are compatible with Θ for all indices i in some index set, then so is the union $\bigcup_i X_i$. The set B of all subsets of $\mathfrak{U}$ that are compatible with Θ is therefore a complete *Boolean* subuniverse of the complex algebra $\mathfrak{Cm}(U)$. The next lemma characterizes when B is in fact a complete subuniverse (and not just a complete Boolean subuniverse) of $\mathfrak{Cm}(U)$.

Lemma 1.21. *A congruence* Θ *on a relational structure* $\mathfrak{U}$ *is bounded if and only if the set of all subsets of* $\mathfrak{U}$ *that are compatible with* Θ *is a complete subuniverse of the complex algebra* $\mathfrak{Cm}(U)$.

Proof. Assume first that the given congruence Θ on $\mathfrak{U}$ is bounded. We have already seen that the set B of all subsets of $\mathfrak{U}$ that are compatible with Θ is a complete Boolean subuniverse of $\mathfrak{Cm}(U)$. It remains to show that B is closed under the operators of $\mathfrak{Cm}(U)$. Focus on the case of a ternary relation R in $\mathfrak{U}$ and the corresponding binary operator $\circ$ in $\mathfrak{Cm}(U)$. Suppose that X and Y are sets in B. The product $X \circ Y$ is defined by

$$X \circ Y = \{w : R(u, v, w) \text{ for some } u \in X \text{ and } v \in Y\}. \tag{1}$$

Let w be any element in $X \circ Y$, and let u and v be elements in X and Y respectively such that $R(u, v, w)$ holds. If $\bar{w} \equiv w \mod \Theta$, then there must be elements $\bar{u}$ and $\bar{v}$ in $\mathfrak{U}$ such that

$$\bar{u} \equiv u \mod \Theta, \qquad \bar{v} \equiv v \mod \Theta, \qquad \text{and} \qquad R(\bar{u}, \bar{v}, \bar{w}), \tag{2}$$

because the congruence Θ is assumed to be bounded (see Definition 1.20). The sets X and Y are compatible with Θ because they are in B, and u belongs to X, and v to Y, so the equivalence classes u/Θ and v/Θ must be included in X and Y respectively, by the definition of compatibility. It follows from the first two equivalences in (2) that that $\bar{u}$ is in X and $\bar{v}$ in Y. Consequently, $\bar{w}$ must be in $X \circ Y$, by (1) and the final part of (2). Conclusion: if an element w belongs to the set $X \circ Y$, then the equivalence class w/Θ is included in $X \circ Y$, so this set is compatible with Θ and therefore belongs to B. Thus, B is a complete subuniverse of $\mathfrak{Cm}(U)$.

To prove the converse direction of the lemma, suppose that B is a complete subuniverse of $\mathfrak{Cm}(U)$, with the goal of showing that the congruence Θ is bounded. Assume u, v, w, and $\bar{w}$ are elements in $\mathfrak{U}$ such that

$$R(u, v, w) \qquad \text{and} \qquad \bar{w} \equiv w \mod \Theta. \tag{3}$$

The sets

$$X = u/\Theta \qquad \text{and} \qquad Y = v/\Theta \tag{4}$$

are obviously compatible with Θ and therefore belong to B, by the definition of B. It follows that the product $X \circ Y$ also belongs to B

and is therefore compatible with Θ, because of the assumption that B is a subuniverse of $\mathfrak{Cm}(U)$. The element w belongs to $X \circ Y$, by the first part of (3) and the definitions in (1) and (4). Consequently, the equivalence class w/Θ is included in the set $X \circ Y$, by the compatibility of $X \circ Y$ with Θ. In particular, the element $\bar{w}$ belongs to $X \circ Y$, by the second part of (3). In view of (1), there must be elements $\bar{u}$ in X and $\bar{v}$ in Y such that $R(\bar{u}, \bar{v}, \bar{w})$ holds. Use (4) to obtain

$$\bar{u} \equiv u \mod \Theta \qquad \text{and} \qquad \bar{v} \equiv v \mod \Theta,$$

as required in Definition 1.20. Thus, Θ is a bounded congruence on $\mathfrak{U}$. $\square$

The following corollary is an immediate consequence of the previous lemma.

Corollary 1.22. *If Θ is a bounded congruence on a relational structure $\mathfrak{U}$, then the set of all sets in $\mathfrak{Cm}(U)$ that are compatible with Θ is a complete subuniverse of $\mathfrak{Cm}(U)$.*

We now reverse the procedure. Begin with an arbitrary complete *Boolean* subuniverse B of the complex algebra $\mathfrak{Cm}(U)$ of a relational structure $\mathfrak{U}$. Thus, B is a subset of the universe of $\mathfrak{Cm}(U)$ that contains the empty set and is closed under the formation of complements and of unions of arbitrary systems of elements in B. A complete (Boolean) field of sets is necessarily atomic (see, for example, Corollary 3 on p. 123, or Lemma 4 on p. 124, of [10]), so B is an atomic Boolean subuniverse of $\mathfrak{Cm}(U)$. In particular, the set of atoms in B is a partition of the universe of $\mathfrak{U}$ in the sense that any two distinct atoms are non-empty and disjoint, and the union of the set of atoms is the universe U. Let Θ be the congruence relation (that is to say, the equivalence relation) on $\mathfrak{U}$ whose equivalence classes are the atoms in B. Every element in B is the union of a set of atoms (because B is atomic), and therefore every element in B is the union of a set of equivalence classes of Θ. Consequently, every element in B is a subset of $\mathfrak{U}$ that is compatible with Θ. Conversely, if X is a subset of $\mathfrak{U}$ that is compatible with Θ, then X must be the union of a set of equivalence classes of Θ; in other words, X must be the union of a set of atoms in B. The set B is assumed to be a complete Boolean subuniverse of $\mathfrak{Cm}(U)$, so B is closed under unions of arbitrary sets of atoms in B. Consequently, the set X must belong to B. The following lemma has been proved.

Lemma 1.23. *If B is a complete Boolean subuniverse of the complex algebra of a relational structure $\mathfrak{U}$, and if Θ is the congruence relation whose equivalence classes are the atoms in B, then B is the set of all subsets of $\mathfrak{U}$ that are compatible with Θ.*

The congruence Θ defined in the preceding lemma is bounded if and only if the complete Boolean subuniverse B is actually a subuniverse of $\mathfrak{Cm}(U)$ in the full sense of the word, by Lemma 1.21. This leads to the following corollary.

Corollary 1.24. *If B is a complete subuniverse of the complex algebra of a relational structure $\mathfrak{U}$, and if Θ is the congruence relation whose equivalence classes are the atoms in B, then Θ is a bounded congruence on $\mathfrak{U}$.*

For a relational structure $\mathfrak{U}$, the complete subuniverse of $\mathfrak{Cm}(U)$ consisting of the sets that are compatible with a bounded congruence Θ on $\mathfrak{U}$ is called the *(first) dual*, or the *dual complete subuniverse*, of Θ, and the bounded congruence Θ whose equivalence classes are the atoms in a complete subuniverse B of $\mathfrak{Cm}(U)$ is called the *(first) dual*, or the *dual bounded congruence*, of B.

If B is the dual of a bounded congruence Θ on $\mathfrak{U}$, then by definition, B consists of the subsets of $\mathfrak{U}$ that are compatible with Θ. Every equivalence class of Θ is compatible with Θ, and these equivalence classes are clearly the minimal non-empty subsets of $\mathfrak{U}$ that are compatible with Θ. Consequently, these equivalence classes must be the atoms in B. The dual of B is, by definition, the bounded congruence Ψ on $\mathfrak{U}$ whose equivalence classes are the atoms in B. Since every equivalence relation is determined by its equivalence classes, and since Θ and Ψ have the same equivalence classes—namely the atoms in B—it follows that $\Theta = \Psi$. Thus, the second dual of every bounded congruence is itself.

If Θ is the dual of a complete subuniverse B of $\mathfrak{Cm}(U)$, then B is the set of elements in $\mathfrak{Cm}(U)$ that are compatible with Θ, by Lemma 1.23. The dual of Θ is, by definition, the complete subuniverse C of $\mathfrak{Cm}(U)$ that consists of the elements that are compatible with Θ. Consequently, $B = C$. Thus, every complete subuniverse of $\mathfrak{Cm}(U)$ is its own second dual.

Suppose Θ and Ψ are bounded congruences on $\mathfrak{U}$, and B and C are the respective dual complete subuniverses of $\mathfrak{Cm}(U)$. If Θ is included in Ψ, then every equivalence class of Ψ is a union of equivalence classes of Θ. Every atom in C is therefore a union of atoms in B, so C is included in B. Conversely, if C is included in B, then every atom in C must be a union of atoms in B, and therefore every equivalence class of Ψ must be a union of equivalence classes of Θ. The congruence Θ is therefore included in Ψ. Conclusion:

$$\Theta \subseteq \Psi \qquad \text{if and only if} \qquad C \subseteq B,$$

so the function mapping every bounded congruence on $\mathfrak{U}$ to its dual complete subuniverse is a dual lattice isomorphism from the lattice of bounded congruences on $\mathfrak{U}$ to the lattice of complete subuniverses of $\mathfrak{Cm}(U)$. The following theorem summarizes these results.

Theorem 1.25. *The dual of every bounded congruence on a relational structure $\mathfrak{U}$ is a complete subuniverse of $\mathfrak{Cm}(U)$, and the dual of every complete subuniverse of $\mathfrak{Cm}(U)$ is a bounded congruence on $\mathfrak{U}$. The second dual of every bounded congruence and of every complete subuniverse is itself. The function mapping each bounded congruence on $\mathfrak{U}$ to its dual complete subuniverse is a dual lattice isomorphism from the lattice of bounded congruences on $\mathfrak{U}$ to the lattice of complete subuniverses of $\mathfrak{Cm}(U)$.*

1.9 Duality for Complete Subalgebras

The duality between bounded congruences and complete subuniverses implies a duality between the corresponding structures. To formulate this duality, it is necessary to define the notion of the quotient of a relational structure modulo a bounded congruence, and to prove the basic results about this notion.

Definition 1.26. The *quotient* of a relational structure $\mathfrak{U}$ modulo a bounded congruence Θ on $\mathfrak{U}$ is the relational structure $\mathfrak{V}$ whose universe is the set

$$V = U/\Theta = \{u/\Theta : u \in U\}$$

of congruence classes modulo Θ of elements in $\mathfrak{U}$, and whose fundamental relations are the natural quotients of the fundamental relations in $\mathfrak{U}$, that is to say, for all fundamental relations R of rank n and all sequences of elements $r_0, \ldots, r_{n-1}$ in V, the relationship $R(r_0, \ldots, r_{n-1})$ holds in $\mathfrak{V}$ if and only if there are elements $u_0, \ldots, u_{n-1}$ in $\mathfrak{U}$ such that $r_i = u_i/\Theta$ for $i < n$ and $R(u_0, \ldots, u_{n-1})$ holds in $\mathfrak{U}$. The function that maps each element u in $\mathfrak{U}$ to its equivalence class u/Θ is called the *quotient mapping* (or the *quotient function*) from $\mathfrak{U}$ onto $\mathfrak{V}$ (*induced by Θ*). □

Every bounded homomorphism ϑ on a relational structure $\mathfrak{U}$ gives rise to a bounded congruence relation on $\mathfrak{U}$, namely the *kernel* of ϑ, which is defined to be the relation Θ on the universe of $\mathfrak{U}$ determined by

$$u \equiv v \mod \Theta \qquad \text{if and only if} \qquad \vartheta(u) = \vartheta(v).$$

Lemma 1.27. *If ϑ is a bounded homomorphism from a relational structure $\mathfrak{U}$ into a relational structure $\mathfrak{V}$, then the kernel of ϑ is a bounded congruence on $\mathfrak{U}$.*

Proof. Let Θ be the kernel of ϑ. It is routine to check that Θ is an equivalence relation (and thus a congruence relation) on the universe of $\mathfrak{U}$. To show that Θ is bounded, focus on the case of a ternary relation R. Consider elements u, v, w, and $\bar{w}$ in $\mathfrak{U}$, and suppose that

$$\bar{w} \equiv w \mod \Theta. \tag{1}$$

Write

$$r = \vartheta(u), \qquad s = \vartheta(v), \qquad t = \vartheta(w). \tag{2}$$

If $R(u, v, w)$ holds in $\mathfrak{U}$, then $R(r, s, t)$ holds in $\mathfrak{V}$, by the homomorphism properties of ϑ. Since

$$t = \vartheta(w) = \vartheta(\bar{w}),$$

by (1) and the definition of Θ, it follows that $R(r, s, \vartheta(\bar{w}))$ holds in $\mathfrak{V}$. The homomorphism ϑ is assumed to be bounded, so there must be elements $\bar{u}$ and $\bar{v}$ in $\mathfrak{U}$ such that

$$\vartheta(\bar{u}) = r, \qquad \vartheta(\bar{v}) = s, \qquad \text{and} \qquad R(\bar{u}, \bar{v}, \bar{w}). \tag{3}$$

From the first two equations in (2) and (3), and the definition of Θ as the kernel of ϑ, it follows that

$$\bar{u} \equiv u \mod \Theta \qquad \text{and} \qquad \bar{v} \equiv v \mod \Theta.$$

The congruence Θ is therefore bounded, by Definition 1.20 and the final part of (3). □

The following analogue for relational structures of the Homomorphism Theorem for algebras says that every bounded congruence Θ is the kernel of a bounded epimorphism, and in fact Θ is the kernel of the corresponding quotient mapping.

Theorem 1.28. *If Θ is a bounded congruence on a relational structure $\mathfrak{U}$, then the corresponding quotient function is an bounded epimorphism from $\mathfrak{U}$ to $\mathfrak{U}/\Theta$ that has Θ as its kernel.*

Proof. Again, focus on the case of a ternary relation R. Obviously, the quotient function—call it ϑ—maps the universe of $\mathfrak{U}$ onto the universe of $\mathfrak{U}/\Theta$. It is easy to check that ϑ preserves the fundamental relations of the relational structures. Indeed, if u, v, and w are elements in $\mathfrak{U}$, and if

$$r = u/\Theta = \vartheta(u), \qquad s = v/\Theta = \vartheta(v), \qquad t = w/\Theta = \vartheta(w),$$

then $R(u, v, w)$ in $\mathfrak{U}$ implies $R(r, s, t)$ in $\mathfrak{U}/\Theta$, by the definition of a quotient relational structure, so ϑ preserves the relation R in the sense that

$$R(u, v, w) \qquad \text{implies} \qquad R(\vartheta(u), \vartheta(v), \vartheta(w)).$$

Thus, ϑ is an epimorphism.

To see that ϑ is bounded, consider elements r and s in the quotient $\mathfrak{U}/\Theta$ and an element $\bar{w}$ in $\mathfrak{U}$, such that $R(r, s, \vartheta(\bar{w}))$ holds in the quotient. The definition of the quotient implies that there must be elements u, v, and w in $\mathfrak{U}$ such that $R(u, v, w)$ and

$$r = u/\Theta, \qquad s = v/\Theta, \qquad \vartheta(\bar{w}) = w/\Theta. \tag{1}$$

Since $\vartheta(\bar{w}) = \bar{w}/\Theta$, by the definition of ϑ, it follows from the final equation in (1) that $\bar{w} \equiv w \mod \Theta$. The congruence Θ is assumed to be bounded, so there must be elements $\bar{u}$ and $\bar{v}$ in $\mathfrak{U}$ such that

$$\bar{u} \equiv u \mod \Theta, \qquad \bar{v} \equiv v \mod \Theta, \qquad \text{and} \qquad R(\bar{u}, \bar{v}, \bar{w}), \tag{2}$$

by Definition 1.20. Use (1) and (2), and the definition of the quotient function ϑ to obtain

$$\vartheta(\bar{u}) = \bar{u}/\Theta = u/\Theta = r \qquad \text{and} \qquad \vartheta(\bar{v}) = \bar{v}/\Theta = v/\Theta = s.$$

The epimorphism ϑ is therefore bounded, by Definition 1.8.

The kernel of ϑ is the congruence Ψ defined by

$$u \equiv v \quad \mod \Psi \qquad \text{if and only if} \qquad \vartheta(u) = \vartheta(v).$$

Since ϑ is the quotient mapping induced by Θ, we have

$$\vartheta(u) = \vartheta(v) \qquad \text{if and only if} \qquad u/\Theta = v/\Theta,$$
$$\text{if and only if} \qquad u \equiv v \quad \mod \Theta.$$

Combine these observations to conclude that $\Psi = \Theta$, and therefore Θ is the kernel of ϑ. $\square$

The next theorem is the analogue for relational structures of the First Isomorphism Theorem for algebras.

Theorem 1.29. *Every bounded homomorphic image of a relational structure $\mathfrak{U}$ is isomorphic to a quotient of $\mathfrak{U}$ modulo a bounded congruence relation. In fact, if ϑ is a bounded epimorphism from $\mathfrak{U}$ to $\mathfrak{V}$, and if Θ is the kernel of ϑ, then the function mapping u/Θ to $\vartheta(u)$ for each element u is an isomorphism from $\mathfrak{U}/\Theta$ to $\mathfrak{V}$.*

Proof. The kernel of the bounded epimorphism ϑ is the relation Θ on $\mathfrak{U}$ defined by

$$u \equiv v \quad \mod \Theta \qquad \text{if and only if} \qquad \vartheta(u) = \vartheta(v), \tag{1}$$

and this relation is a bounded congruence on $\mathfrak{U}$, by Lemma 1.27. It is clear from (1) that the function ϱ from $\mathfrak{U}/\Theta$ to $\mathfrak{V}$ defined by

$$\varrho(u/\Theta) = \vartheta(u)$$

is a well-defined bijection from the universe of $\mathfrak{U}/\Theta$ to the universe of $\mathfrak{V}$.

To check that ϱ is an isomorphism, focus on the case of a ternary relation R. Consider arbitrary elements r, s, and t in the quotient structure $\mathfrak{U}/\Theta$. Assume first that

$$R(r, s, t) \tag{2}$$

holds in the quotient. In this case, there must be elements u, v, and w in $\mathfrak{U}$ such that

$$R(u, v, w) \tag{3}$$

holds in $\mathfrak{U}$, and

$$r = u/\Theta, \qquad s = v/\Theta, \qquad t = w/\Theta, \tag{4}$$

by the definition of the quotient structure. From (4) and the definition of ϱ, we obtain

$$\varrho(r) = \varrho(u/\Theta) = \vartheta(u), \qquad \varrho(s) = \varrho(v/\Theta) = \vartheta(v), \tag{5}$$
$$\varrho(t) = \varrho(w/\Theta) = \vartheta(w).$$

The function ϑ is assumed to be a homomorphism, so

$$R(\vartheta(u), \vartheta(v), \vartheta(w)) \tag{6}$$

in $\mathfrak{V}$, by (3) and Definition 1.8, and therefore

$$R(\varrho(r), \varrho(s), \varrho(t)), \tag{7}$$

in $\mathfrak{V}$, by (5).

Assume now that (7) holds in $\mathfrak{V}$, with the goal of deriving (2). Let u, v, and w be elements in $\mathfrak{U}$ such that (4) holds. The equations in (5) must hold, by (4) and the definition of ϱ, so (6) must hold, by (5) and (7). The epimorphism ϑ is assumed to be bounded, so there must be elements $\bar{u}$ and $\bar{v}$ in $\mathfrak{U}$ such that

$$\vartheta(\bar{u}) = \vartheta(u), \qquad \vartheta(\bar{v}) = \vartheta(v), \qquad \text{and} \qquad R(\bar{u}, \bar{v}, w), \tag{8}$$

by (6) and Definition 1.8. From (8), the definition of the relation Θ in (1), and the definition of the quotient structure $\mathfrak{U}/\Theta$, we obtain

$$\bar{u} \equiv u \mod \Theta, \qquad \bar{v} \equiv v \mod \Theta, \qquad R(\bar{u}/\Theta, \bar{v}/\Theta, w/\Theta). \tag{9}$$

From the first parts of (4) and (9) we get

$$r = u/\Theta = \bar{u}/\Theta \qquad \text{and} \qquad s = v/\Theta = \bar{v}/\Theta.$$

Since $t = w/\Theta$, by the last equation in (4), the preceding equations and the last part of (9) lead to the conclusion that (2) holds. Thus, ϑ is an isomorphism. $\square$

We are now ready to formulate and prove the duality theorem for complete subalgebras and quotient relational structures.

Theorem 1.30. *Suppose* Θ *is a bounded congruence on a relational structure* $\mathfrak{U}$, *and* $\mathfrak{B}$ *is the complete subalgebra of* $\mathfrak{Cm}(U)$ *whose universe is the dual of* Θ. *The dual relational structure of* $\mathfrak{B}$ *is the quotient structure* $\mathfrak{U}/\Theta$, *and the dual algebra of* $\mathfrak{U}/\Theta$—*that is to say, the complex algebra* $\mathfrak{Cm}(U/\Theta)$—*is isomorphic to* $\mathfrak{B}$.

Proof. The dual relational structure of $\mathfrak{B}$ is the atom structure $\mathfrak{V}$ of $\mathfrak{B}$. The elements in $\mathfrak{V}$ are the atoms in $\mathfrak{B}$, and $\mathfrak{B}$ is a complete subalgebra of $\mathfrak{Cm}(U)$, so the elements in $\mathfrak{V}$ are non-empty disjoint sets that have the universe of $\mathfrak{U}$ as their union. The complex algebra $\mathfrak{Cm}(V)$ is isomorphic to $\mathfrak{B}$ via the function ψ that maps each subset of $\mathfrak{V}$—that is to say, each set of atoms in $\mathfrak{B}$—to its supremum in $\mathfrak{B}$, by Theorem 1.3. Since $\mathfrak{B}$ is a complete subalgebra of $\mathfrak{Cm}(U)$, the supremum of a set of atoms is just the union of the set, and the function ψ is a complete monomorphism from $\mathfrak{Cm}(V)$ into $\mathfrak{Cm}(U)$. In particular, for each atom r in $\mathfrak{V}$, we have

$$\psi(\{r\}) = \bigcup\{r\} = r. \tag{1}$$

The dual of ψ is the bounded epimorphism ϑ from $\mathfrak{U}$ to $\mathfrak{V}$ that is defined by

$$\vartheta(u) = r \quad \text{if and only if} \quad u \in \psi(\{r\}),$$

by Theorem 1.10. In view of (1), this definition may be written as

$$\vartheta(u) = r \quad \text{if and only if} \quad u \in r.$$

In other words, ϑ maps each element in $\mathfrak{U}$ to the element in $\mathfrak{V}$—that is to say, to the atom in $\mathfrak{B}$—that contains u.

The kernel of ϑ is the congruence Ψ on $\mathfrak{U}$ that is defined by

$$u \equiv v \mod \Psi \quad \text{if and only if} \quad \vartheta(u) = \vartheta(v).$$

In other words, u is congruent to v modulo Ψ if and only if u and v belong to the same atom in $\mathfrak{B}$. The universe of $\mathfrak{B}$ is assumed to be the dual of the given bounded congruence Θ, and therefore Θ is the dual of the universe of $\mathfrak{B}$, by Theorem 1.25. Consequently, Θ consists of those pairs (u, v) such that u and v belong to the same atom in $\mathfrak{B}$, by the definition of the dual of a complete subuniverse. It follows that $\Psi = \Theta$, and therefore Θ is the kernel of ϑ.

In view of Theorem 1.29 and the conclusion of the preceding paragraph, the quotient $\mathfrak{U}/\Theta$ is isomorphic to $\mathfrak{V}$ via the canonical mapping that takes each congruence class u/Θ to the element $\vartheta(u)$. The congruence Θ is the dual of the universe of $\mathfrak{B}$, so the congruence classes of Θ are just the atoms in $\mathfrak{B}$, by the definition of the dual of a complete subuniverse. In particular, u/Θ is the atom in $\mathfrak{B}$ to which u belongs. The element $\vartheta(u)$ is also the atom in $\mathfrak{B}$ to which u belongs, by the observations made above, so $\vartheta(u) = u/\Theta$. It follows that the canonical isomorphism from $\mathfrak{U}/\Theta$ to $\mathfrak{V}$, which maps u/Θ to $\vartheta(u)$, is just the identity function on the universe of $\mathfrak{U}/\Theta$. Consequently,

$$\mathfrak{U}/\Theta = \mathfrak{V},$$

so $\mathfrak{U}/\Theta$ is the atom structure of $\mathfrak{B}$.

The dual of the atom structure $\mathfrak{U}/\Theta$ is, by definition, the complex algebra $\mathfrak{Cm}(U/\Theta)$. Since $\mathfrak{B}$ is isomorphic to the complex algebra of its atom structure, by Theorem 1.3, it follows from the conclusion of the preceding paragraph that $\mathfrak{B}$ is isomorphic to $\mathfrak{Cm}(U/\Theta)$. □

As Mai Gehrke has pointed out, there are weaker, less explicit versions of Theorems 1.19 and 1.30 that are well known from the literature and that follow almost immediately from the epi-mono duality between bounded homomorphisms and complete homomorphisms formulated in Theorem 1.11. To describe these weaker versions, consider two relational structures $\mathfrak{U}$ and $\mathfrak{V}$.

If $\mathfrak{V}$ is isomorphic to an inner substructure of $\mathfrak{U}$, say via a bounded monomorphism ϑ, then the dual of ϑ is a complete epimorphism φ from $\mathfrak{Cm}(U)$ to $\mathfrak{Cm}(V)$, so that $\mathfrak{Cm}(V)$ is a complete homomorphic image of $\mathfrak{Cm}(U)$. Conversely, if $\mathfrak{Cm}(V)$ is a complete homomorphic image of $\mathfrak{Cm}(U)$, say via a complete epimorphism φ, then the dual of φ is a bounded monomorphism ϑ from $\mathfrak{V}$ to $\mathfrak{U}$, so that $\mathfrak{V}$ is isomorphic to an inner substructure of $\mathfrak{U}$. Thus, there is a duality between (isomorphic copies of) inner substructures of $\mathfrak{U}$ and complete homomorphic images of $\mathfrak{Cm}(U)$.

In a similar vein, if $\mathfrak{V}$ is a bounded homomorphic image of $\mathfrak{U}$, say via a bounded epimorphism ϑ, then the dual of ϑ is a complete monomorphism φ from $\mathfrak{Cm}(V)$ to $\mathfrak{Cm}(U)$, so that $\mathfrak{Cm}(V)$ is isomorphic to a complete subalgebra of $\mathfrak{Cm}(U)$. Conversely, if $\mathfrak{Cm}(V)$ is isomorphic to a complete subalgebra of $\mathfrak{Cm}(U)$, say via a complete monomorphism φ, then the dual of φ is a bounded epimorphism from $\mathfrak{U}$ to $\mathfrak{V}$, so that $\mathfrak{V}$ is a bounded homomorphic image of $\mathfrak{U}$. Thus, there is a duality between bounded homomorphic images of $\mathfrak{U}$ and (isomorphic copies of) complete subalgebras of $\mathfrak{Cm}(U)$.

For relational structures with a single binary relation (frames) and complete and atomic Boolean algebras with a single complete unary operator (complete and atomic modal algebras), the dualities just described are formulated in Corollary 1 of Thomason [39]. For more general similarity types, the observation that if $\mathfrak{V}$ is an inner substructure of $\mathfrak{U}$, then $\mathfrak{Cm}(V)$ is a homomorphic image of $\mathfrak{Cm}(U)$ (without the assertion of completeness), and the observation that if $\mathfrak{V}$ is a bounded homomorphic image of $\mathfrak{U}$, then $\mathfrak{Cm}(V)$ is isomorphic to a subalgebra of $\mathfrak{Cm}(U)$ (without the assertion of completeness), occur as parts (1) and (2) of Corollary 3.2.5 in Goldblatt [13]. (See also Propositions 5.51 and 5.52, and Theorem 5.47, in Blackburn-de Rijke-Venema [1], as well as Theorems 5.8 and 5.37 in Venema [40].)

Theorems 1.19 and 1.30 strengthen these known duality results by making explicit exactly how the dual structures are constructed. First of all, a duality between complete ideals and inner subuniverses is constructed—the dual of a complete ideal M is the inner subuniverse V that is the complement of the principal generator of M, and the dual of an inner subuniverse V is the complete ideal M generated by the complement of V—and this duality proves to be a dual lattice isomorphism (Theorem 1.18). Similarly, a duality between bounded congruences and complete subuniverses is constructed—the dual of a bounded congruence Θ is the complete subuniverse B that consists of the unions of the various sets of congruence classes of Θ, and the dual of a complete subuniverse B is the bounded congruence Θ whose congruence classes are the atoms in B—and this duality also proves to be a dual lattice isomorphism (Theorem 1.25).

Return now to the relational structures $\mathfrak{U}$ and $\mathfrak{V}$. If $\mathfrak{V}$ is isomorphic to an inner substructure, say $\mathfrak{W}$, of $\mathfrak{U}$, then the dual algebra of $\mathfrak{V}$, namely the complex algebra $\mathfrak{Cm}(V)$, is a complete homomorphic image of $\mathfrak{Cm}(U)$, and in fact it is isomorphic to the quotient of $\mathfrak{Cm}(U)$ modulo the complete ideal that is the dual of the universe of the inner substructure $\mathfrak{W}$. Conversely, if $\mathfrak{Cm}(V)$ is a complete homomorphic image of $\mathfrak{Cm}(U)$, say via a complete epimorphism φ, then $\mathfrak{Cm}(V)$ is isomorphic to the quotient of $\mathfrak{Cm}(U)$ modulo the complete ideal M that is the kernel of φ, and therefore the dual structure $\mathfrak{V}$ is isomorphic to the inner substructure of $\mathfrak{U}$ whose universe is the dual of the complete ideal M. This is the content of Theorem 1.19.

Similarly, if $\mathfrak{V}$ is a bounded homomorphic image of $\mathfrak{U}$, say via a bounded epimorphism ϑ, then $\mathfrak{V}$ is isomorphic to the quotient of $\mathfrak{U}$ modulo the bounded congruence Θ that is the kernel of ϑ, and therefore the dual algebra $\mathfrak{Cm}(V)$ is isomorphic to the complete subalgebra of $\mathfrak{Cm}(U)$ whose universe is the dual of the bounded congruence Θ. Conversely, if $\mathfrak{Cm}(V)$ is isomorphic to a complete subalgebra, say $\mathfrak{B}$, of $\mathfrak{Cm}(U)$, then the dual structure $\mathfrak{V}$ is isomorphic to the quotient of $\mathfrak{U}$ modulo the bounded congruence that is the dual of the universe of the complete subalgebra $\mathfrak{B}$, so that $\mathfrak{V}$ is a bounded homomorphic image of $\mathfrak{U}$. This is the content of Theorem 1.30.

1.10 Duality for Products

The duality between relational structures and complex algebras carries with it a duality between disjoint unions of relational structures and direct products of complex algebras. Here, the term *disjoint* means that the universes of the structures are mutually disjoint.

Definition 1.31. The *union* of a system $(\mathfrak{U}_i : i \in I)$ of mutually disjoint relational structures $\mathfrak{U}_i$ is defined to be the relational structure $\mathfrak{U}$ in which the universe is the (disjoint) union, over all indices i, of the universes of the structures $\mathfrak{U}_i$, and each fundamental relation in $\mathfrak{U}$ is the (disjoint) union, over all indices i, of the corresponding fundamental relations in the structures $\mathfrak{U}_i$. The relational structures $\mathfrak{U}_i$ are called the *components* of the union $\mathfrak{U}$.

The *disjoint union* of an arbitrary system $(\mathfrak{V}_i : i \in I)$ of relational structures is defined to be the union of the disjoint system $(\mathfrak{U}_i : i \in I)$ in which $\mathfrak{U}_i$ is the isomorphic image of the relational structure $\mathfrak{V}_i$ under the function that maps each element r in $\mathfrak{V}_i$ to the element (r, i). □

The second half of this definition means that in order to create the union of an arbitrary system $(\mathfrak{V}_i : i \in I)$ of relational structures, we first create an isomorphic copy $\mathfrak{U}_i$ of each relational structure $\mathfrak{V}_i$ in such a way that the resulting system $(\mathfrak{U}_i : i \in I)$ of relational structures is disjoint, and we then form the union of this disjoint system. In the particular version of this construction given in the above definition, the universe of $\mathfrak{U}_i$ is the set of elements

$$U_i = \{(r, i) : r \in V_i\},$$

and the fundamental relations in $\mathfrak{U}_i$ are defined to be the images of
the fundamental relations in $\mathfrak{V}_i$ under the function that maps each
element r in $\mathfrak{V}_i$ to the pair (r, i). In other words, a fundamental relation
of rank n in $\mathfrak{U}_i$ is defined to hold for a sequence $((r_0, i), \dots, (r_{n-1}, i))$
of elements from $\mathfrak{U}_i$ just in case the corresponding relation in $\mathfrak{V}_i$ holds
for the corresponding sequence $(r_0, \dots, r_{n-1})$ of elements from $\mathfrak{V}_i$.

The notions of the external and internal direct products of systems
of Boolean algebras with operators are defined in the standard way.
By the *external (direct) product* of a system of Boolean algebras with
operators $(\mathfrak{A}_i : i \in I)$, we understand the algebra $\mathfrak{A}$ in which the
universe is the system of functions r with domain I such that $r(i)$
is in $\mathfrak{A}_i$ for each index i, and in which the operations are defined
coordinatewise. For example, if f is an operator of rank n, then

$$f(r_0, \dots, r_{n-1}) = (f(r_0(i), \dots, r_{n-1}(i)) : i \in I)$$

for each sequence $r_0, \dots, r_{n-1}$ of elements in $\mathfrak{A}$, and similarly for the
operations of addition and complement. (The operation f on the left
side of this equation is the one being defined in $\mathfrak{A}$, while the ones on
the right are the operators of the factor algebras $\mathfrak{A}_i$.) The notation

$$\mathfrak{A} = \prod_i \mathfrak{A}_i$$

is used to denote the external product.

The definition of the notion of an internal product is well known
from group theory, but less well known in the more general context of
abstract algebras (see, for example, Jónsson-Tarski [20]). We shall only
need a special case of this definition. Consider a system $(\mathfrak{U}_i : i \in I)$
of disjoint relational structures, and for each i, suppose that $\mathfrak{A}_i$ is a
subalgebra of the complex algebra $\mathfrak{Cm}(U_i)$. By the *internal (direct)
product* of the system $(\mathfrak{A}_i : i \in I)$, we understand the algebra $\mathfrak{A}$ in
which the universe consists of the sets of the form $X = \bigcup_i X_i$, where X_i
is an element in $\mathfrak{A}_i$ (and hence a subset of U_i) for each i in I, and in
which the operations are defined as follows: if

$$X = \bigcup_i X_i \qquad \text{and} \qquad Y = \bigcup_i Y_i$$

are elements in $\mathfrak{A}$, then

$$X + Y = \bigcup_i (X_i + Y_i) \qquad \text{and} \qquad -X = \bigcup_i -X_i,$$

and if f is an operator of rank n, and if

$$X_0 = \bigcup_i X_{0,i}, \qquad \dots \qquad, \quad X_{n-1} = \bigcup_i X_{n-1,i}$$

is a sequence of elements in $\mathfrak{A}$, then

$$f(X_0, \dots, X_{n-1}) = \bigcup_i f(X_{0,i}, \dots, X_{n-1,i}).$$

(The operations on the left sides of the preceding equations are those being defined in $\mathfrak{A}$, while the ones on the right sides, after the union symbol, are the operations of the algebras $\mathfrak{A}_i$.) Obviously, the internal product of the system is canonically isomorphic to the external product of the system via the function that maps each element $\bigcup_i X_i$ in the internal product to the element $(X_i : i \in I)$ in the external product.

Suppose $(\mathfrak{U}_i : i \in I)$ is a disjoint system of relational structures, and $\mathfrak{U}$ is the union of this system. It turns out that the internal product $\mathfrak{A}$ of the system $(\mathfrak{Cm}(U_i) : i \in I)$ of complex algebras coincides with the complex algebra $\mathfrak{Cm}(U)$. For the proof, observe first that the universe of $\mathfrak{U}_i$ is included in the universe of $\mathfrak{U}$, and $\mathfrak{Cm}(U)$ is closed under arbitrary unions, so every element in $\mathfrak{A}$ is certainly an element in $\mathfrak{Cm}(U)$. The converse is also true: every element in $\mathfrak{Cm}(U)$ belongs to $\mathfrak{A}$. In fact, since the sets U_i are assumed to be mutually disjoint and have U as their union, every element X in $\mathfrak{Cm}(U)$ can be represented in exactly one way as a union $X = \bigcup_i X_i$, where X_i is a subset of U_i for each index i, namely the subset $X_i = X \cap U_i$. Thus, the universes of $\mathfrak{A}$ and $\mathfrak{Cm}(U)$ are the same.

To see that the operations of the two algebras are also the same, suppose

$$X = \bigcup_i X_i \qquad \text{and} \qquad Y = \bigcup_i Y_i$$

are the unique representations of two sets X and Y in $\mathfrak{Cm}(U)$. The sum of X and Y in $\mathfrak{Cm}(U)$ is, by definition, the union of these two sets, and for each index i, the sum of the sets X_i and Y_i in $\mathfrak{Cm}(U_i)$ is the union of these two sets, so the operation of addition in $\mathfrak{Cm}(U)$ must be the same as it is in $\mathfrak{A}$. In more detail,

$$X + Y = X \cup Y = \left(\bigcup_i X_i\right) \cup \left(\bigcup_i Y_i\right) = \bigcup_i (X_i \cup Y_i) = \bigcup_i (X_i + Y_i)$$

(where the left-most addition is performed in $\mathfrak{Cm}(U)$, and the right-most addition is performed in $\mathfrak{Cm}(U_i)$), so the sum $X + Y$ is the same in $\mathfrak{Cm}(U)$ as it is in $\mathfrak{A}$. A similar argument, using the disjointness of the sets U_i, shows that the complement $-X$ is the same in $\mathfrak{Cm}(U)$ as it is in $\mathfrak{A}$.

To check that the operators in the two algebras are the same, consider the case of a binary operator $\circ$ that has been defined in terms of a ternary relation R. The product $X \circ Y$ in $\mathfrak{Cm}(U)$ is, by definition, the set of elements t such that $R(r, s, t)$ holds in $\mathfrak{U}$ for some r in X and s in Y, and the product $X_i \circ Y_i$ in $\mathfrak{Cm}(U_i)$ is the sets of elements t such that $R(r, s, t)$ holds in $\mathfrak{Cm}(U_i)$ for some r in X_i and some s in Y_i. The relation R in $\mathfrak{U}$ is defined to be the union of the corresponding relations in $\mathfrak{Cm}(U_i)$, so if t belongs to $X \circ Y$, and if r and s are elements in X and Y respectively such that $R(r, s, t)$ holds in $\mathfrak{U}$, then there must be an index i such that $R(r, s, t)$ holds in $\mathfrak{U}_i$. In this case, r and s must belong to the sets

$$X \cap U_i = X_i \qquad \text{and} \qquad Y \cap U_i = Y_i$$

respectively, so t belongs to the product $X_i \circ Y_i$ (formed in $\mathfrak{Cm}(U_i)$). Conversely, if t belongs to one of the products $X_i \circ Y_i$, then $R(r, s, t)$ holds in $\mathfrak{U}_i$ for some r in X_i and s in Y_i. The elements r and s of course belong to the sets X and Y respectively, by the definitions of these sets, and $R(r, s, t)$ holds in $\mathfrak{U}$, by the definition of the relation R in $\mathfrak{U}$, so t must belong to the product $X \circ Y$. Thus,

$$X \circ Y = \bigcup_i X_i \circ Y_i$$

(where the product on the left is formed in $\mathfrak{Cm}(U)$, and those on the right are formed in $\mathfrak{Cm}(U_i)$), and this is precisely the definition of $X \circ Y$ in $\mathfrak{A}$. Conclusion: $\mathfrak{A} = \mathfrak{Cm}(U)$. The following theorem has been proved.

Theorem 1.32. *The internal product of a system of complex algebras of mutually disjoint relational structures is equal to the complex algebra of the union of the relational structures.*

Using the preceding theorem, it is easy to arrive at a general conclusion about products of complex algebras.

Corollary 1.33. *The external product of an arbitrary system of complex algebras of relational structures is canonically isomorphic to the complex algebra of the disjoint union of the relational structures.*

The corollary says that the dual relational structure of the direct product of a system of complex algebras is isomorphic to the disjoint union of the dual relational structures of the complex algebras. Put somewhat differently, the dual algebra of the disjoint union of a system of relational structures is isomorphic to the direct product of the corresponding system of dual algebras of the relational structures. Theorem 1.32 says that the dual algebra of the union is in fact *equal* to the internal product of the corresponding system of dual algebras.

The special case of Corollary 1.33 in which the relational structures have a single binary relation (frames) is formulated in Corollary 1 of Thomason [39] and in Theorem 6.5 of Goldblatt [12]. The general result for arbitrary similarity types is formulated in Lemma 3.4.1 of Goldblatt [13]. (See also Theorem 5.48 in Blackburn-de Rijke-Venema [1], as well as part (v) of Theorem 5.37 in Venema [40].)

1.11 Duality for Ultraproducts

The duality between disjoint unions of relational structures and products of the dual algebras has a parallel duality for ultraproducts. Before developing this duality, we need to introduce the notion of a completion of a Boolean algebra with operators.

Definition 1.34. A Boolean algebra with operators $\mathfrak{B}$ is called a *completion* of a Boolean algebra with operators $\mathfrak{A}$ if the following conditions are satisfied.

  (i) $\mathfrak{B}$ is complete, and $\mathfrak{A}$ is a subalgebra of $\mathfrak{B}$.
 (ii) Every non-zero element in $\mathfrak{B}$ is above a non-zero element in $\mathfrak{A}$.
(iii) For every operator f of rank n and every sequence of elements $r_0, \ldots, r_{n-1}$ in $\mathfrak{B}$, we have

$$f(r_0, \ldots, r_{n-1})$$
$$= \sum \{ f(t_0, \ldots, t_{n-1}) : t_i \in A \text{ and } t_i \leq r_i \text{ for } i < n \}. \ \square$$

In the case of a binary operator $\circ$, the third condition assumes the form
$$r \circ s = \sum \{ p \circ q : p, q \in A \text{ and } p \leq r \text{ and } q \leq s \}.$$

It is worth pointing out that the requirement in condition (i) that $\mathfrak{B}$ be complete means, by definition, that in particular the operators in $\mathfrak{B}$ are complete.

It was shown by Monk [27] that every Boolean algebra with complete operators $\mathfrak{A}$ has a completion and that any two completions of $\mathfrak{A}$ are isomorphic via a mapping that is the identity function on $\mathfrak{A}$. Consequently, one may speak of *the* completion of $\mathfrak{A}$. The algebra $\mathfrak{A}$ and its completion have exactly the same atoms. From this and condition (ii) in Definition 1.34 it follows, in particular, that the completion of $\mathfrak{A}$ is atomic if and only if $\mathfrak{A}$ is itself atomic.

We now return to the question of duality for ultraproducts. The notion of an ultraproduct of a system of algebraic structures is well known from model theory. In order to formulate the definition for a system of Boolean algebras with operators, it is helpful to employ some terminology and notation. Consider a non-empty system $(A_i : i \in I)$ of sets (that is to say, the index set I is not empty) and write $A = \prod_i A_i$ for the Cartesian product of this system. If D is an ultrafilter on the index set I, then an equivalence relation $\equiv_D$ on the product set A can be defined by

$$r \equiv_D s \qquad \text{if and only if} \qquad \{i \in I : r(i) = s(i)\} \in D$$

for all elements r and s in A. Write r/D for the equivalence class of an element r modulo D (or, more precisely, modulo the relation $\equiv_D$).

Let $(\mathfrak{A}_i : i \in I)$ be a non-empty system of Boolean algebras with operators, and write $\mathfrak{A} = \prod_i \mathfrak{A}_i$ for the external product of the system. The *ultraproduct* of the system *modulo* an ultrafilter D on I is defined to be the algebra with universe

$$A/D = \{r/D : r \in A\},$$

in which each operator f of rank n is determined by

$$f(r_0/D, \ldots, r_{n-1}/D) = f(r_0, \ldots, r_{n-1})/D$$

for every sequence of elements $r_0, \ldots, r_{n-1}$ in $\mathfrak{A}$, and similarly for the operations of addition and complement. (The operation f on the left side of this equation is the one being defined on A/D, while the one on the right is the operator in the external product $\mathfrak{A}$.) This ultraproduct is denoted by $(\prod_i \mathfrak{A}_i)/D$, or more simply, by $\mathfrak{A}/D$, when it is understood that $\mathfrak{A}$ is the external product of the given system of Boolean algebras with operators. Notice that there is a matter to be checked here, namely that the definitions of the operations in $\mathfrak{A}/D$ do not depend on the specific representatives of the elements in $\mathfrak{A}/D$ that are being used.

Ultraproducts of systems of relational structures are defined in a completely analogous manner. In more detail, the *ultraproduct* of a non-empty system $(\mathfrak{U}_i : i \in I)$ of relational structures *modulo* an ultrafilter D on I is defined to be the relational structure with universe

$$U/D = \{r/D : r \in U\},$$

where $U = \prod_i U_i$, in which each fundamental relation R of rank n is determined to hold for a sequence of elements $r_0/D, \ldots, r_{n-1}/D$ in U/D just in case the set

$$\{i \in I : R(r_0(i), \ldots, r_{n-1}(i)) \text{ holds in } \mathfrak{U}_i\}$$

belongs to the ultrafilter D. Again, it must be checked that this definition does not depend on the specific representatives of the elements that are being used. The same notation is used to denote ultraproducts of systems of relational structures as is used to denote ultraproducts of algebras.

The general notion of an ultraproduct is due to Łoś [25], although the formulation he uses is very different from the one given above. The formulation we have used is due to Frayne, Morel, and Scott [6]. The basic result about ultraproducts is the Fundamental Theorem of Ultraproducts, due originally to Łoś [25] and, in its present formulation, to Frayne, Morel, and Scott [6]. It says that a formula of first-order logic with n free variables is satisfied by a sequence of elements $r_0/D, \ldots, r_{n-1}/D$ in an ultraproduct $(\prod_i \mathfrak{A}_i)/D$ if and only if the ultrafilter D contains the set of all indices i for which the sequence of elements $r_0(i), \ldots, r_{n-1}(i)$ satisfies the same formula in $\mathfrak{A}_i$. In particular, a sentence of first-oder logic is true in the ultraproduct if and only if the ultrafilter D contains the set of indices i for which that sentence is true in $\mathfrak{A}_i$.

The next theorem—the duality theorem for ultraproducts—connects ultraproducts of complex algebras with ultraproducts of relational structures.

Theorem 1.35. *Let $(\mathfrak{U}_i : i \in I)$ be a non-empty system of non-empty relational structures, and $\mathfrak{U}$ the direct product of this system. For every ultrafilter D on the set I, the completion of the ultraproduct $(\prod_i \mathfrak{Cm}(U_i))/D$ of the complex algebras is isomorphic to the complex algebra $\mathfrak{Cm}(U/D)$ of the ultraproduct $\mathfrak{U}/D$.*

Proof. Write $\mathfrak{A} = \prod_i \mathfrak{Cm}(U_i)$, and let D be an ultrafilter on the index set I. Each of the factor complex algebras $\mathfrak{Cm}(U_i)$ is atomic, and the property of being atomic is preserved under the passage to ultraproducts, by the Fundamental Theorem of Ultraproducts. Consequently, the ultraproduct $\mathfrak{A}/D$ is atomic, and in fact, the atoms are the elements of the form p/D, where p is an element in $\mathfrak{A}$ such that the set

$$\{i \in I : p(i) \text{ is an atom in } \mathfrak{Cm}(U_i)\}$$

belongs to D. The atoms in $\mathfrak{Cm}(U_i)$ are really singletons of elements in $\mathfrak{U}_i$, but these singletons may be identified with the elements themselves. Under this identification, one can view each atom in $\mathfrak{A}/D$ as the quotient modulo D of an element p in $\mathfrak{A}$ such that the set

$$I_p = \{i \in I : p(i) \in U_i\}$$

belongs to D. The convention of identifying each element with its singleton also identifies each atom in the complex algebra $\mathfrak{Cm}(U/D)$ with the element in $\mathfrak{U}/D$ of which it is the singleton.

The intuition underlying the proof of the theorem is that each atom in $\mathfrak{A}/D$ can be identified with an element in $\mathfrak{U}/D$, and vice versa. Under this identification, the set of atoms in $\mathfrak{A}/D$ coincides with the set of elements in $\mathfrak{U}/D$, which in turn coincides with the set of atoms in $\mathfrak{Cm}(U/D)$.

The way to make this intuition precise is to define a correspondence φ from the set of atoms in $\mathfrak{A}/D$ to the set U/D. Write

$$\varphi(p/D) = r/D$$

whenever p/D is an atom in $\mathfrak{A}/D$ and r is an element in $\mathfrak{U}$ that agrees with p on the set I_p. Notice that the notations p/D and r/D, though visually similar, are being used here in two different ways: p/D is the set of elements in $\mathfrak{A}$ that agree with p on a set in D, whereas r/D is the set of elements in $\mathfrak{U}$ that agree with r on a set in D. By virtue of the identification of elements with their singletons, every element r in $\mathfrak{U}$ may be viewed as an element in $\mathfrak{A}$, that is to say, $(r(i) : i \in I)$ is identified with $(\{r(i)\} : i \in I)$. Nevertheless, when forming quotients modulo D, the equivalence class r/D in $\mathfrak{A}/D$ is not the same as the equivalence class r/D in $\mathfrak{U}/D$. Rather, the former includes the latter as a proper subset.

Straightforward arguments show that φ is well defined and one-to-one. Indeed, let r and s be elements in $\mathfrak{U}$, and p/D and q/D atoms in $\mathfrak{A}/D$ such that r agrees with p on the set I_p, and s agrees with q on the set I_q. If $p/D = q/D$, then the set I_1 of indices on which p and q agree belongs to the ultrafilter D, by the definition of the relation $\equiv_D$. Since the sets I_p and I_q also belong to D, by the assumption that p/D and q/D are atoms, the intersection $I_1 \cap I_p \cap I_q$ must be in D. The set I_2 of indices on which r and s agree includes this intersection, because

$$r(i) = p(i) = q(i) = s(i)$$

for each index i in the intersection. Therefore, I_2 is in D, so $r/D = s/D$, by the definition of $\equiv_D$. Consequently, φ is well defined. Conversely, if $r/D = s/D$, then the set I_2 belongs to D, and therefore so does the intersection $I_2 \cap I_p \cap I_q$. The set I_1 includes this intersection and is therefore in D, so $p/D = q/D$. Consequently, φ is one-to-one.

To see that the domain of φ is the set of all atoms in $\mathfrak{A}/D$, consider an arbitrary atom p/D in this set. Each of the atom structures $\mathfrak{U}_i$ is non-empty, by assumption, so for each index i we may fix an element q_i in $\mathfrak{U}_i$. Define an element r in $\mathfrak{U}$ by

$$r(i) = \begin{cases} p(i) & \text{if } i \in I_p, \\ q_i & \text{if } i \notin I_p. \end{cases}$$

Since r agrees with p on the set I_p, we have $\varphi(p/D) = r/D$, and therefore p/D is in the domain of φ. To see that the range of φ is all of U/D, observe that every element r in $\mathfrak{U}$ may be consider to belong to $\mathfrak{A}$, by virtue of the identification of elements with their singletons, and $\varphi(r/D) = r/D$, by the definition of φ (where the first quotient is formed in $\mathfrak{A}/D$ and the second in $\mathfrak{U}/D$). Conclusion: φ is a well-defined bijection from the set of atoms in $\mathfrak{A}/D$ to the universe of $\mathfrak{U}/D$.

The argument in the preceding paragraph also implies that the atoms in $\mathfrak{A}/D$ are precisely the quotients in $\mathfrak{A}/D$ of elements in $\mathfrak{U}$. We may therefore conceive of φ as the function defined by $\varphi(r/D) = r/D$ for every r in $\mathfrak{U}$, provided we keep in mind that the two quotients involved in this definition do not denote the same set; the first quotient is formed in $\mathfrak{A}/D$ and the second in $\mathfrak{U}/D$.

Under the identification of elements with their singletons, the universe of $\mathfrak{U}/D$ coincides with the set of atoms in the complex algebra $\mathfrak{Cm}(U/D)$. The mapping φ may therefore be viewed as a bijection from the set of atoms in $\mathfrak{A}/D$ to the set of atoms in $\mathfrak{Cm}(U/D)$. The next step is to prove that φ satisfies the condition formulated in Theorem 1.1. Focus on the case of a binary operator $\circ$ that is defined in terms of a ternary relation R. Let r, s, and t be elements in $\mathfrak{U}$. The inequality

$$t/D \leq r/D \circ s/D \tag{1}$$

holds in $\mathfrak{A}/D$ if and only if the set

$$\{i \in I : t(i) \leq r(i) \circ s(i) \text{ holds in } \mathfrak{Cm}(U_i)\}$$

belongs to D, by the Fundamental Theorem of Ultraproducts. Use the definition of the operator $\circ$ in $\mathfrak{Cm}(U_i)$ to see that the latter set belongs to D if and only if the set

$$\{i \in I : R(r(i), s(i), t(i)) \text{ holds in } \mathfrak{U}_i\} \tag{2}$$

belongs to D. On the other hand, the inequality (1) holds in $\mathfrak{Cm}(U/D)$ if and only if

$$R(r/D, s/D, t/D)$$

holds in $\mathfrak{U}/D$, by the definition of the operator $\circ$ in $\mathfrak{Cm}(U/D)$. Use the definition of the relation R in $\mathfrak{U}/D$ to see that the latter holds if and only if the set (2) belongs to D. Combine these observations to conclude that the inequality

$$t/D \leq r/D \circ s/D$$

holds in $\mathfrak{A}/D$ if and only if it holds in $\mathfrak{Cm}(U/D)$ (with the quotients being formed in $\mathfrak{A}/D$ and in $\mathfrak{U}/D$ respectively), that is to say,

$$t/D \leq r/D \circ s/D \quad \text{if and only if} \quad \varphi(t/D) \leq \varphi(r/D) \circ \varphi(s/D).$$

Let $\mathfrak{B}$ be the completion of the ultraproduct $\mathfrak{A}/D$. This completion is a complete and atomic Boolean algebra with operators, it has the same atoms as $\mathfrak{A}/D$, and the values of the operations in $\mathfrak{B}$ on atoms coincide with the values of the corresponding operations in $\mathfrak{A}/D$ on atoms. It follows that φ is a bijection from the set of atoms in $\mathfrak{B}$ to the set of atoms in $\mathfrak{Cm}(U/D)$, and the equivalences in the preceding paragraph continue to hold. Apply Theorem 1.1 to conclude that φ can be extended to an isomorphism from $\mathfrak{B}$ to $\mathfrak{Cm}(U/D)$. $\square$

The preceding theorem says that the dual algebra of an ultraproduct of a system of relational structures is, up to isomorphism, the completion of the corresponding ultraproduct of the system of corresponding dual algebras.

Monk [26] proved that an ultraproduct of a system of relation algebras that are complex algebras of projective geometries is embeddable into the complex algebra of the corresponding ultraproduct of the

projective geometries. Goldblatt extended Monk's result by proving
(in Lemma 3.6.5 of [13]) that an ultraproduct of a system of complex algebras of relational structures is embeddable into the complex
algebra of the ultraproduct of the corresponding system of the relational structures. Ian Hodkinson has kindly called our attention to
the following result related to Theorem 1.35 and stated in Lemma 3.7
of Gehrke, Harding, and Venema [8]: if L is an ultraproduct of a system $(L_i : i \in I)$ of bounded lattices with additional monotone operations (all of the same similarity type), then the β-MacNeille completion
of L is isomorphic to the β-MacNeille completion of the corresponding
ultraproduct of the system of completions of the lattice expansions L_i.

Chapter 2
Topological Duality

In this chapter we study the topological duality that exists between relational spaces—that is to say, relational structures endowed with the topology of a Boolean space—and arbitrary Boolean algebras with operators. The duality between the spaces and algebras carries with it a corresponding duality between morphisms: every continuous bounded homomorphism between relational spaces corresponds to a homomorphism between the dual Boolean algebras with operators, and conversely. The duality between the morphisms implies other dualities as well. Here are some examples. Every special open subset of a relational space corresponds to an ideal in the dual algebra, and vice versa. Every relational congruence on a relational space corresponds to a subuniverse of the dual algebra, and vice versa. The equivalence classes of compactifications of the disjoint union of a system of relational spaces correspond to the subalgebras of the direct product of the system of dual algebras that include the weak direct product of the system of dual algebras, and vice versa. In particular, the Stone-Čech compactification of a disjoint union of relational spaces corresponds to the direct product of the dual algebras.

2.1 Topological Duality for Boolean Algebras

To motivate the development that follows, we recall some of the essential features of the topological duality for Boolean algebras that was discovered by Stone [36], [37]. Correlated with every Boolean algebra A is a certain topological space U. The points in the space U are the ultrafilters in A. The clopen subsets of U are the sets of the form

S. Givant, *Duality Theories for Boolean Algebras with Operators*,
Springer Monographs in Mathematics, DOI 10.1007/978-3-319-06743-8_2,
© Springer International Publishing Switzerland 2014

$$F_r = \{X \in U : r \in X\}$$

for elements r in A, and the open subsets are the unions of arbitrary systems of clopen sets. Under this topology, U becomes a Boolean space, that is to say, it becomes a compact Hausdorff space in which the clopen sets (the sets that are simultaneously closed and open) form a base for the topology. The space U is called the *dual space* of the algebra A.

Inversely, correlated with every Boolean space is a certain Boolean algebra A. The elements in A are the clopen subsets of U, and the Boolean operations are the set-theoretic operations of union and complement. This Boolean algebra is called the *dual algebra* of the space U.

If one starts with a Boolean algebra A, forms its dual space U, and then forms the dual algebra of U, then the result is the Boolean algebra B of sets of the form F_r for r in A. The algebras A and B are isomorphic via the function that maps every element r in A to the clopen set F_r. Similarly, if one starts with a Boolean space U, forms its dual algebra A, and then forms the dual space of A, the result is the Boolean space V in which the points are the ultrafilters of elements in A, that is to say, the ultrafilters of clopen subsets of U; these ultrafilters have the form

$$X_r = \{F \in A : r \in F\}$$

for elements r in $\mathfrak{U}$. The spaces U and V are homeomorphic via the function that maps each point r in U to the ultrafilter X_r. This whole state of affairs may be expressed by saying that each Boolean algebra is isomorphic to its second dual and each Boolean space is homeomorphic to its second dual.

The duality between Boolean algebras and Boolean spaces is accompanied by a corresponding duality—apparently due to Sikorski (see [35])—between the homomorphisms on the algebras and the continuous functions on the spaces. Let U and V be Boolean spaces, and A and B the corresponding dual Boolean algebras. If ϑ is a continuous function from U into V, then there is a natural Boolean homomorphism φ from B into A that is defined by

$$\varphi(F) = \vartheta^{-1}(F) = \{u \in U : \vartheta(u) \in F\}$$

for elements F in B, that is to say, for clopen subsets F of V. This mapping is called the (*first*) *dual* of the function ϑ. Inversely, if φ is a homomorphism from B into A, then there is a natural continuous function ϑ from U to V that is defined by

$$\vartheta(u) = r \qquad \text{if and only if} \qquad r \in \bigcap \{F \in B : u \in \varphi(F)\}.$$

for elements r in U. This function is called the (*first*) *dual* of the homomorphism φ. If one starts with a homomorphism φ from B to A, forms the dual continuous mapping ϑ from U to V, and then forms the dual homomorphism of ϑ, the result is the original homomorphism φ. Similarly, if one starts with a continuous function ϑ from U to V, forms the dual homomorphism φ from B to A, and then forms the dual continuous mapping of φ, the result is the original function ϑ. This whole state of affairs may be expressed by saying that every continuous function between Boolean spaces, and every homomorphism between Boolean algebras, is its own second dual. Furthermore, a continuous mapping between Boolean spaces is one-to-one or onto if and only if its dual homomorphism is onto or one-to-one respectively.

Finally, if ϑ is a continuous function from U to V, and δ a continuous function from V to a Boolean space W, and if φ and ψ are the respective dual homomorphisms of ϑ and δ, then the dual of the composition $\delta \circ \vartheta$ is just the composition $\varphi \circ \psi$. The category of Boolean algebras with homomorphisms as morphisms is therefore dually equivalent to the category of Boolean spaces with continuous functions as morphisms.

The epi-mono duality between morphisms implies a duality between ideals and open sets, and a duality between subuniverses and Boolean congruences; there is a corresponding duality between quotient algebras and subspaces on the one hand, and a corresponding duality between subalgebras and quotient spaces on the other hand. The epi-mono duality also implies a duality between certain subdirect products of Boolean algebras and compactifications of unions of Boolean spaces, and in particular between direct products of Boolean algebras and Stone-Čech compactifications of unions of Boolean spaces. (See, for example, [10], Chapters 34–38, 43, where references to the literature may also be found; or see [23], Chapter 3.)

We shall see that completely analogous dualities hold for Boolean algebras with operators and relational spaces. Only a very basic knowledge of topology is needed to understand much of this duality, but the duality between subdirect products of algebras and compactifications of unions of spaces requires a slightly deeper knowledge of topology.

2.2 Relational Spaces

We begin with the task of defining what is meant by a topological
relational structure, or, for short, a relational space. When defining
a topological algebra such as a topological group, there are usually
some requirements on the fundamental operations—such as continuity
or openness—that render these operations compatible with the given
topology on the universe of the algebra. An operation of rank n in a
topological algebra is defined to be continuous if the inverse image,
under the operation, of every open set is open in the product topology
on the nth Cartesian power of the universe; and the operation is defined to be open if the image, under the operation, of every open set in
the nth Cartesian power of the universe is open in the universe. A relational structure does not have fundamental operations, but rather
fundamental relations, so one must clarify what it means for a fundamental relation to be continuous or open. The first task is to clarify
what is meant by the image and the inverse image of a set under such
a relation.

The definition of the image of a set under a relation is quite natural
and straightforward. If R is a relation of rank $n + 1$ in a relational
structure $\mathfrak{U}$, and if H is a subset of the nth Cartesian power U^n of the
universe U, then the *image of H under the relation R* is the set

$$R^*(H) = \{t \in U : R(r_0, \ldots, r_{n-1}, t)$$
$$\text{for some sequence } (r_0, \ldots, r_{n-1}) \in H\}.$$

In particular, if $F_0, \ldots, F_{n-1}$ are subsets of U, then

$$R^*(F_0 \times \cdots \times F_{n-1}) = \{t \in U : R(r_0, \ldots, r_{n-1}, t)$$
$$\text{for some } r_i \in F_i \text{ for } i < n\}.$$

In the case of a sequence of n elements $r_0, \ldots, r_{n-1}$ from U, we shall
often write simply $R^*(r_0, \ldots, r_{n-1})$ instead of $R^*(\{(r_0, \ldots, r_{n-1})\})$.

To illustrate these ideas more concrete, consider the case of a ternary
relation R. If H a subset of $U \times U$, then

$$R^*(H) = \{t \in U : R(r, s, t) \text{ for some } (r, s) \in H\},$$

and if F and G are subsets of U, then

$$R^*(F \times G) = \{t \in U : R(r, s, t) \text{ for some } r \in F \text{ and } s \in G\}.$$

The definition of the inverse image of a set under a relation is a bit more involved, because there are several possibilities. The one that works in the present context is the following. If H is a subset of U, then the *inverse image of H under R* is the set

$$R^{-1}(H) = \{(r_0, \ldots, r_{n-1}) \in U^n : R(r_0, \ldots, r_{n-1}, t) \text{ implies } t \in H\}.$$

For example, if R is a ternary relation, then

$$R^{-1}(H) = \{(r, s) \in U \times U : R(r, s, t) \text{ implies } t \in H\}.$$

Warning: this definition of inverse image requires the image of the singleton set

$$\{(r_0, \ldots, r_{n-1})\} = \{r_0\} \times \cdots \times \{r_{n-1}\}$$

under R to be *entirely include* in H, and not just to have a non-empty intersection with H, before the sequence $(r_0, \ldots, r_{n-1})$ is place in the inverse image, so that

$$(r_0, \ldots, r_{n-1}) \in R^{-1}(H) \quad \text{if and only if} \quad R^*(r_0, \ldots, r_{n-1}) \subseteq H.$$

Lemma 2.1. *If R is a relation of rank n on a set U, and if H and K are subsets of U, then*

$$H \subseteq K \qquad \text{implies} \qquad R^{-1}(H) \subseteq R^{-1}(K).$$

Proof. Assume that H is included in K, and let $(r_0, \ldots, r_{n-1})$ be a sequence in $R^{-1}(H)$. The image set $R^*(r_0, \ldots, r_{n-1})$ of this sequence is included in H, by the observation preceding the lemma, and therefore the image set of the sequence is also included in K, by the assumption. Consequently, $(r_0, \ldots, r_{n-1})$ belongs to $R^{-1}(K)$, by the observation preceding the lemma. $\square$

One can imagine a definition of the inverse image of H under R in which a sequence $(r_0, \ldots, r_{n-1})$ is put into $R^{-1}(H)$ just in case there exists an element t in H such that $R(r_0, \ldots, r_{n-1}, t)$ holds. In fact, this is the definition that is adopted by Halmos [15] for binary relations. However, this is not the definition that works in the present context.

With the preceding notions in hand, we can now define what it means for a relation to be open, clopen, or continuous. A relation R of rank $n+1$ is defined to be *open* if the image under R of every open subset of U^n (in the product topology) is an open subset of U, and *clopen* if the image of every clopen subset of U^n is a clopen subset of U. Note that a unary relation is defined to be open or clopen if it is an open subset or a clopen subset respectively of U. A relation R of rank $n + 1$ is defined to be *continuous* if the inverse image of every open subset of U is an open subset of U^n (in the product topology).

Definition 2.2. A *topological relational structure*, or a *relational space* for short, is a relational structure $\mathfrak{U}$, together with a topology on the universe U, such that U is a Boolean space under this topology—that is to say, U is a compact Hausdorff space in which the clopen sets form a base for the topology—and the relations in $\mathfrak{U}$ are clopen, and the relations of rank at least two are continuous. $\square$

Notice that a relation of rank 1 in $\mathfrak{U}$ is always continuous, since the inverse of every subset of U under such a relation is, by definition, a subset of the set $U^0 = \{\varnothing\}$ and is therefore automatically open. Consequently, the final part of the preceding definition amounts to saying that all the relations in $\mathfrak{U}$ are clopen and continuous. In speaking about relational spaces, we shall often use topological terminology. For instance, we may call the elements in the universe "points", and we may use phrases such as "the set F is open in $\mathfrak{U}$" or "the set F is an open subset of $\mathfrak{U}$" to express that F is an open subset of the topological space U.

There have been several earlier definitions of an analogue of a relational space. Halmos [15] is concerned with Boolean algebras with a single unary operator. The corresponding topological structures he considers are Boolean spaces with a single binary relation, say R, that is *Boolean* in the sense that the inverse image (in his sense of the word) of every clopen set under R is clopen, and the image of every point under R is closed (see p. 232 in [15]). Goldblatt [13] is concerned not only with arbitrary Boolean algebras with operators, but more generally with bounded distributive lattices with operators. If we restrict our attention to the relational structures in his paper that correspond to Boolean algebras with operators, then a relational space is for him a relational structure with the topology of a Boolean space and with the property that the inverse images of points under the fundamental relations are closed subsets (in the product topology)—that is to say, the set

$$\{(r_0, \dots, r_{n-1}) : R(r_0, \dots, r_{n-1}, t)\}$$

is closed for each fundamental relation R of rank $n + 1$ and each point t—and the images of clopen subsets (in the product topology) are clopen sets (see his definition of an ordered relational space on pp. 184–185 in [13]). We shall have more to say about the connection between Goldblatt's definition and our own later on. Hansoul [16] is also concerned with bounded distributive lattices with operators. He deals with multi-algebras instead of relational structures, but it seems that in his definition of the appropriate topological multi-algebra, he requires the inverse images of points to be closed, and the images of clopen sets to be clopen (see Definition 1.6 in [16]). Sambin and Vaccaro [32] are concerned with Boolean algebras with a single unary operator that is distributive, not over addition, but rather over multiplication. They consider relational structures that consist of a single binary relation, and they define such a relation R to be continuous if, for every clopen set F, the set of elements r such that $R(r, t)$ implies that t is in F is always clopen.

2.3 Duality for Algebras

The next step is to correlate with each Boolean algebra with operators $\mathfrak{A}$ a relational space $\mathfrak{U}$. The notion of a canonical extension of a Boolean algebra with operators is implicitly involved in this discussion.

Definition 2.3. A *canonical extension* of a Boolean algebra with operators $\mathfrak{A}$ is a Boolean algebra with operators $\mathfrak{B}$ that has the following properties.

  (i) $\mathfrak{B}$ is complete (in particular, the operators in $\mathfrak{B}$ are complete) and atomic, and $\mathfrak{A}$ is a subalgebra of $\mathfrak{B}$.

 (ii) For any two distinct atoms r and s in $\mathfrak{B}$, there is an element t in $\mathfrak{A}$ that *separates* the atoms in the sense that r is below t and s is below $-t$.

(iii) If a subset X of $\mathfrak{A}$ has the supremum 1 in $\mathfrak{B}$, then there is a finite subset Y of X such that 1 is the supremum of Y (in $\mathfrak{A}$ and in $\mathfrak{B}$). $\square$

Condition (ii) is called the *atom separation property*, while condition (iii) is called the *compactness property*. The preceding definition goes back to Jónsson-Tarski [21], where it is shown that every Boolean algebra with operators $\mathfrak{A}$ (even one that is not normal) has a canonical extension, and that any two canonical extensions of $\mathfrak{A}$ are isomorphic via a mapping that is the identity function on $\mathfrak{A}$. This uniqueness up to isomorphisms justifies speaking of *the* canonical extension of $\mathfrak{A}$.

The precise definition of (an isomorphic copy of) the canonical extension plays an important role in the subsequent discussion, so we give some of the details of the construction. Let U be the set of all ultrafilters in $\mathfrak{A}$. With each operator f of rank n in $\mathfrak{A}$, correlate a relation R of rank $n+1$ on U by defining

$$R(X_0, \ldots, X_{n-1}, Z) \qquad \text{if and only if}$$
$$\{f(r_0, \ldots, r_{n-1}) : r_i \in X_i \text{ for } i < n\} \subseteq Z$$

for all ultrafilters $X_0, \ldots, X_{n-1}, Z$ in U. If one extends f in the usual way to a *complex operation* of rank n on subsets of $\mathfrak{A}$, by

$$f(X_0, \ldots X_{n-1}) = \{f(r_0, \ldots, r_{n-1}) : r_i \in X_i \text{ for } i < n\},$$

then the definition of the relation R on sequences of elements in $\mathfrak{U}$ assumes the form

$$R(X_0, \ldots, X_{n-1}, Z) \qquad \text{if and only if} \qquad f(X_0, \ldots X_{n-1}) \subseteq Z.$$

For an operator of rank 0, that is to say, for a distinguished constant c, the definition of R takes the form

$$R(Z) \qquad \text{if and only if} \qquad c \in Z.$$

In the case of a binary operator $\circ$ in $\mathfrak{A}$, the relation R is the ternary relation on the set of ultrafilters that is defined on all triples of ultrafilters X, Y, Z by

$$R(X, Y, Z) \qquad \text{if and only if} \qquad X \circ Y \subseteq Z,$$

where $X \circ Y$ is the *complex product* of X and Y, that is to say,

$$X \circ Y = \{r \circ s : r \in X \text{ and } s \in Y\}$$

Let $\mathfrak{U}$ be the relational structure whose universe is the set U of ultrafilters in $\mathfrak{A}$, and whose fundamental relations are the relations on U that are correlated with the operators in $\mathfrak{A}$. The complex algebra $\mathfrak{Cm}(U)$ is a complete and atomic Boolean algebra with operators, and the sum of every set of elements in $\mathfrak{Cm}(U)$ is just the union of that set of elements. For each element r in $\mathfrak{A}$, let F_r be the set of ultrafilters in $\mathfrak{U}$ that contain the element r, and notice that F_r is a subset of $\mathfrak{U}$ and therefore an element in the complex algebra $\mathfrak{Cm}(U)$. Using properties of ultrafilters, one can check that the following equations hold in $\mathfrak{Cm}(U)$. (The occurrence of f on the right side of the equation in (iii) denotes an arbitrary operator in $\mathfrak{A}$, and the occurrence on the left denotes the corresponding operator in $\mathfrak{Cm}(U)$.)

Lemma 2.4. *Let r, s, t, and $r_0, \ldots, r_{n-1}$ be elements in $\mathfrak{A}$.*

(i) *$F_r = F_s$ if and only if $r = s$.*
(ii) *$F_r \cup F_s = F_t$ if and only if $t = r + s$.*
(iii) *$\sim F_r = F_t$ if and only if $t = -r$.*
(iv) *$f(F_{r_0}, \ldots, F_{r_{n-1}}) = F_t$ if and only if $t = f(r_0, \ldots, r_{n-1})$.*

Proof. The implication from right to left in (i) is trivial. The argument establishing the reverse implication proceeds by contraposition. Suppose r and s are distinct. One of the products $r \cdot -s$ and $-r \cdot s$ is then non-zero, say it is $r \cdot -s$. The set consisting of this single element trivially satisfies the finite meet property, so there must be an ultrafilter Z in $\mathfrak{A}$ that contains $r \cdot -s$, and therefore also r and $-s$, by the upward closure of Z. It follows that Z contains r but not s, so Z belongs to the set F_r but not to F_s. These two sets are therefore distinct.

The verifications of (ii) and (iii) use the basic properties of ultrafilters and are not difficult. For example, to establish the implication from right to left in (ii), suppose that $t = r + s$. The element t belongs to an arbitrary ultrafilter Z (in U) if and only if at least one of the summands r and s is in Z, so

$$
\begin{array}{llll}
Z \in F_r \cup F_s & \text{if and only if} & Z \in F_r \text{ or } Z \in F_s, \\
& \text{if and only if} & r \in Z \text{ or } s \in Z, \\
& \text{if and only if} & t \in Z, \\
& \text{if and only if} & Z \in F_t,
\end{array}
$$

by the definition of the union of two sets, the definitions of the sets F_r, F_s, and F_t, and the above mentioned property of ultrafilters. Thus, the first equation in (ii) holds. To establish the reverse implication, assume that the first equation in (ii) holds. Write $p = r + s$, and observe that

$$F_p = F_r \cup F_s,$$

by the implication just established. Combine this with the assumed equation in (ii) to arrive at $F_p = F_t$. Apply (i) to conclude that $p = t$. This establishes the equivalence in (ii).

The equivalence in (iii) follows by a similar argument, using the property that a complement $-r$ belongs to an ultrafilter Z if and only if r is not in Z.

The proof of the equivalence in (iv) is similar in character to the preceding arguments, but substantially more complicated in details. Consider the case of a binary operator $\circ$. Let r and s be arbitrary elements in $\mathfrak{A}$, and assume first that $t = r \circ s$, with the goal of showing that

$$F_r \circ F_s = F_t. \tag{1}$$

The definition of the operator $\circ$ in $\mathfrak{Cm}(U)$, and the definitions of the sets F_r and F_s imply that

$$F_r \circ F_s = \{Z \in U : X \circ Y \subseteq Z$$
$$\text{for some } X, Y \in U \text{ with } r \in X \text{ and } s \in Y\}. \tag{2}$$

Consequently, if Z belongs to the left side of (1), then there must be ultrafilters X and Y in U containing r and s respectively such that the complex product $X \circ Y$ is included in Z, by (2). The product $r \circ s$ belongs to $X \circ Y$, by the definition of the complex product of two sets, so $r \circ s$ must belong to Z and therefore Z must belong to the right side of (1) by the assumption on t and the definition of the set F_t. This proves that the left side of (1) is included in the right side.

To establish the reverse inclusion, suppose Z belongs to the right side of (1). To show that Z belongs to the left side of (1), ultrafilters X and Y containing r and s respectively must be constructed with the property that $X \circ Y$ is included in Z, by (2). The Boolean dual of Z is the maximal Boolean ideal

$$-Z = \{-r : r \in Z\}$$

in $\mathfrak{A}$. Write

$$W_0 = \{v \in A : r \circ (v \cdot s) \in -Z\}. \tag{3}$$

We proceed to show that W_0 is a proper Boolean ideal containing the element $-s$.

First of all,

$$r \circ (-s \cdot s) = r \circ 0 = 0,$$

by Boolean algebra and the assumption that the operators in $\mathfrak{A}$ are normal. The element 0 belongs to the ideal $-Z$, so $r \circ (-s \cdot s)$ belongs to $-Z$, and consequently $-s$ is in W_0, by (3). In particular, W_0 is not empty.

Second, if u is in W_0, and if v is an element in $\mathfrak{A}$ that is below u, then

$$r \circ (v \cdot s) \leq r \circ (u \cdot s),$$

by Boolean algebra and the monotony of the operators in $\mathfrak{A}$. Since the element $r \circ (u \cdot s)$ is in $-Z$, by (3) (with u in place of v), and since $-Z$ is an ideal (and therefore downward closed), the element $r \circ (v \cdot s)$ must also be in $-Z$, and therefore v must be in W_0, by (3). Thus, W_0 is downward closed.

Third, if u and v are in W_0, then the elements

$$r \circ (u \cdot s) \quad \text{and} \quad r \circ (v \cdot s)$$

must be in $-Z$, by (3). The sum of these two elements is also in $-Z$, because $-Z$ is an ideal and hence closed under addition. Since

$$r \circ [(u + v) \cdot s] = r \circ (u \cdot s) + r \circ (v \cdot s),$$

by Boolean algebra and the distributivity of the operators in $\mathfrak{A}$, it follows that $r \circ [(u + v) \cdot s]$ is in $-Z$, and therefore $u + v$ is in W_0, by (3). Thus, W_0 is closed under addition.

Finally, W_0 does not contain the unit 1. Indeed, the product $r \circ s$ is in Z, by the assumption that Z is in F_t and the assumption about t, so this product cannot be in $-Z$. Since

$$r \circ (1 \cdot s) = r \circ s,$$

by Boolean algebra, it follows that $r \circ (1 \cdot s)$ is not in $-Z$, and consequently 1 is not in W_0, by (3). This completes the proof that W_0 is a proper Boolean ideal containing $-s$.

Use the Maximal Ideal Theorem for Boolean algebras (see Section 1.2) to extend W_0 to a maximal Boolean ideal W in $\mathfrak{A}$. The Boolean dual of W is the ultrafilter Y that is determined by

$$Y = -W = \{-v : v \in W\} = \{w \in A : w \notin W\} = A \sim W. \quad (4)$$

The element $-s$ is in W_0 and therefore also in W, so s is in Y, by (4). Also, if v is Y, then v is not in W, by (4), and therefore v is not in W_0.

Consequently, the element $r \circ (v \cdot s)$ cannot be in $-Z$, by (3), so this element must be in Z. Since

$$r \circ (v \cdot s) \leq r \circ v,$$

by Boolean algebra and the monotony of the operators in $\mathfrak{A}$, and since Z is a filter, it may be concluded that $r \circ v$ is in Z, by the upward closure of filters. Thus,

$$\{r \circ v : v \in Y\} \subseteq Z. \tag{5}$$

Define a subset V_0 of $\mathfrak{A}$ by

$$V_0 = \{u \in A : (u \cdot r) \circ v \in -Z \text{ for some } v \in Y\}. \tag{6}$$

As before, the set V_0 is a proper Boolean ideal in $\mathfrak{A}$ that contains the element $-r$. For example, $-r$ is in V_0, by (6), because 0 is in the ideal $-Z$, and 1 is in Y, and

$$(-r \cdot r) \circ 1 = 0 \circ 1 = 0,$$

by Boolean algebra and the assumption that $\circ$ is a normal operator.

The argument that V_0 is downward closed is similar to the argument that the set W_0 is downward closed, and is left to the reader. To see that V_0 is closed under addition, assume u and v are in V_0. There must be elements w_1 and w_2 in Y such that

$$(u \cdot r) \circ w_1 \qquad \text{and} \qquad (v \cdot r) \circ w_2 \tag{7}$$

are in $-Z$, by (6). Write $w = w_1 \cdot w_2$, and observe that w is in Y, by the closure of filters under multiplication. The elements

$$(u \cdot r) \circ w \qquad \text{and} \qquad (v \cdot r) \circ w \tag{8}$$

are respectively below the elements in (7), by Boolean algebra and the monotony of the operators in $\mathfrak{A}$, so they, too, belong to the ideal $-Z$, by the downward closure of ideals. The sum of the elements in (8) is therefore also in $-Z$, by the closure of ideals under addition. Since

$$[(u + v) \cdot r] \circ w = (u \cdot r) \circ w + (v \cdot r) \circ w,$$

by Boolean algebra and the distributivity of the operator $\circ$, it follows from the preceding observations and (6) that $u + v$ belongs to V_0. Thus, V_0 is an ideal.

To see that the ideal V_0 is proper, observe that

$$(1 \cdot r) \circ v = r \circ v,$$

by Boolean algebra, and that $r \circ v$ belongs to Z for every v in Y, by (5). Consequently, $(1 \cdot r) \circ v$ cannot be in $-Z$ for any v in Y, so 1 cannot be in V_0, by (6).

Extend V_0 to a maximal Boolean ideal V. The Boolean dual of V is the ultrafilter

$$X = -V = \{-u : u \in V\} = \{w \in A : w \notin V\} = A \sim V. \qquad (9)$$

The element $-r$ is in V_0, and V_0 is included in V, so $-r$ must be in V, and therefore r must be in X, by (9). If u is any element in X, then u cannot be in V, by (9), and therefore u cannot be in V_0. Consequently, the element $(u \cdot r) \circ v$ cannot belong to $-Z$ for any v in Y, by (6), so this element must be in Z for every such v. Since Z is a filter, and therefore upward closed, and since

$$(u \cdot r) \circ v \leq u \circ v,$$

it may be concluded that $u \circ v$ belongs to Z for every v in Y. This conclusion holds for all u in X, so the complex product $X \circ Y$ is included in Z, as desired.

This completes the proof of (1) and of the implication from right to left in (iv). The proof of the reverse implication in (iv) is nearly identical to the proof of the corresponding implication in (ii) (with $+$ and $F_r \cup F_s$ replaced by $\circ$ and $F_r \circ F_s$ respectively) and is left to the reader. □

The equivalences in Lemma 2.4 imply that the set B of all subsets of $\mathfrak{U}$ of the form F_r, for r in $\mathfrak{A}$, is closed under the operations of $\mathfrak{Cm}(U)$ and is therefore subuniverse of $\mathfrak{Cm}(U)$. Let $\mathfrak{B}$ be the corresponding subalgebra. The equivalences also imply that the function φ mapping each element r in $\mathfrak{A}$ to the set F_r of ultrafilters containing r is an isomorphism from $\mathfrak{A}$ to $\mathfrak{B}$. It is called the *canonical embedding* of $\mathfrak{A}$ into $\mathfrak{Cm}(U)$.

We proceed to show that $\mathfrak{Cm}(U)$ is the canonical extension of $\mathfrak{B}$. The canonical extension of $\mathfrak{A}$ can then be obtained by applying the Exchange Principle from general algebra to the pair $\mathfrak{B}$ and $\mathfrak{Cm}(U)$, exchanging $\mathfrak{B}$ for $\mathfrak{A}$, and $\mathfrak{Cm}(U)$ for an isomorphic copy of $\mathfrak{Cm}(U)$ that is the desired canonical extension of $\mathfrak{A}$. As regards the atom separation property, observe that the atoms in $\mathfrak{Cm}(U)$ are the singletons of ultrafilters. For any two distinct ultrafilters X and Y, there must be an element r that belongs to one of the ultrafilters but not the other, say r is in X but not in Y. The complement $-r$ belongs to Y (because Y is an ultrafilter that does not contain r), so the atom $\{X\}$ is included in F_r, while the atom $\{Y\}$ is included in F_{-r}. Thus, the atom separation property holds for $\mathfrak{Cm}(U)$ with respect to $\mathfrak{B}$.

To verify the compactness property, argue by contraposition. Consider an arbitrary subset G of $\mathfrak{B}$ with the property that no finite subset of G has the unit $F_1 = U$ in $\mathfrak{Cm}(U)$ as its union. It is to be shown that G cannot have U as its union. The set G must have the form

$$G = \{F_r : r \in I\}$$

for some subset I of $\mathfrak{A}$, by the definition of $\mathfrak{B}$. The assumed property of G and the isomorphism φ from $\mathfrak{A}$ to $\mathfrak{B}$ imply that no finite subset of I has 1 as its sum in $\mathfrak{A}$. Consequently, the product of every finite subset of the set

$$-I = \{-r : r \in I\}$$

is different from 0, by Boolean algebra, that is to say, the set $-I$ has the finite meet property. It follows that the set $-I$ can be extended to an ultrafilter X. Since the complement of every element in I belongs to X, none of the elements r in I can belong to X, and therefore X cannot belong to any of the sets F_r in G. It follows that the union of the sets in G does not contain all of the ultrafilters in U, since it does not contain X. Consequently, the union of G is not the unit U.

We turn now to the problem of correlating a relational space with $\mathfrak{A}$. Let $\mathfrak{U}$ be the relational structure defined above, consisting of the ultrafilters in $\mathfrak{A}$, and let $\mathfrak{B}$ be the subalgebra of $\mathfrak{Cm}(U)$ that is the isomorphic image of $\mathfrak{A}$ under the canonical embedding. There is a natural topology that can be defined on the universe of $\mathfrak{U}$ with the help of the algebra $\mathfrak{B}$. The universe of $\mathfrak{B}$—that is to say, the collection of sets of the form F_r for elements r in $\mathfrak{A}$—is declared to be a base of the topology, so that the open sets in the topology are the unions of arbitrary systems of sets in $\mathfrak{B}$. Using the fact that $\mathfrak{B}$ is closed under intersection and contains the sets $\varnothing = F_0$ and $U = F_1$, it is easy to check that this class of open sets really does satisfy the three conditions for being a topology, namely it contains the empty set and the universal set, it is closed under unions of arbitrary systems of open sets, and it is closed under intersections of finite systems of open sets. We shall refer to this topology as the *topology induced on* $\mathfrak{U}$ *by* $\mathfrak{A}$ (because the topology is completely determined by the universe of $\mathfrak{A}$ and the ultrafilters in $\mathfrak{A}$). Since the algebra $\mathfrak{B}$ is closed under the formation of complements, the sets in the base of the topology are both open and closed; they constitute the *clopen sets* of the topology. We shall return to this point in a moment.

Under this topology, the space $\mathfrak{U}$ is Hausdorff, and in fact any two points in $\mathfrak{U}$ are separated by a clopen set in $\mathfrak{B}$. This is just a topological manifestation of the atom separation property for the canonical extension: distinct points in $\mathfrak{U}$ are distinct ultrafilters X and Y in $\mathfrak{A}$, and for any two such ultrafilters, the atom separation property of $\mathfrak{Cm}(U)$ with respect to $\mathfrak{B}$ requires the existence of an element r in $\mathfrak{A}$ such that X belongs to F_r, but Y does not (see above). The clopen set F_r therefore separates the points X and Y.

The space $\mathfrak{U}$ is also compact, and in fact, compactness is a topological manifestation of the compactness property of the canonical extension. In order to see this, suppose a system of open sets covers the space $\mathfrak{U}$. Each open set in the system is a union of clopen sets, by the definition of the topology, so there must be a system $(F_r : r \in I)$ of clopen sets (where I is some subset of $\mathfrak{A}$) such that each of these clopen sets is contained in one of the open sets in the given open cover, and the union of this system of clopen sets is U. The compactness property for $\mathfrak{Cm}(U)$ with respect to $\mathfrak{B}$ implies the existence of a finite subset I_0 of I such that the union of the system $(F_r : r \in I_0)$ is U. Since each of the clopen sets in this finite subsystem is included in one of the open sets in the given open cover, a finite subsystem of the original open cover must also have U as its union.

It is well known that in a compact topological space, if a given Boolean algebra of clopen sets separates points, then the space is in fact a Boolean space, and the given Boolean algebra of clopen sets is in fact the Boolean algebra of all clopen subsets of the space (see Lemma 1 on p. 305 of [10]). It follows that the topology defined on $\mathfrak{U}$ turns this relational structure into a Boolean space, and the clopen subsets of $\mathfrak{U}$ are precisely the sets of the form F_r for r in $\mathfrak{A}$. In order to show that $\mathfrak{U}$ is in fact a relational space, it remains to prove that the fundamental relations in $\mathfrak{U}$ are clopen, and that the fundamental relations of rank at least two are continuous.

Lemma 2.5. *The fundamental relations in the relational structure* $\mathfrak{U}$ *correlated with a Boolean algebra with operators* $\mathfrak{A}$ *are clopen.*

Proof. Focus on the case of a ternary relation R in $\mathfrak{U}$ and the corresponding binary operator $\circ$ in $\mathfrak{Cm}(U)$. Consider first the special case of the product two clopen subsets of $\mathfrak{U}$, say F_r and F_s. It is to be shown that the image of the set $F_r \times F_s$ under the relation R is clopen. This image is defined to be the set

$$R^*(F_r \times F_s) = \{Z \in U : R(X, Y, Z)$$
$$\text{for some } X \in F_r \text{ and } Y \in F_s\}. \quad (1)$$

The sets F_r and F_s both belong to the complex algebra $\mathfrak{Cm}(U)$, so it is possible to form their product $F_r \circ F_s$ in $\mathfrak{Cm}(U)$. Use the definition of $\circ$ in $\mathfrak{Cm}(U)$ and (1) to obtain

$$F_r \circ F_s = \{Z \in U : R(X, Y, Z)$$
$$\text{for some } X \in F_r \text{ and } Y \in F_s\} = R^*(F_r \times F_s).$$

If $t = r \circ s$ in $\mathfrak{A}$, then $F_t = F_r \circ F_s$ in $\mathfrak{Cm}(U)$, by Lemma 2.4(iv), so the preceding computation shows that

$$R^*(F_r \times F_s) = F_r \circ F_s = F_t. \quad (2)$$

Since F_t is a clopen subset of $\mathfrak{U}$, by the definition of the topology on $\mathfrak{U}$, the image of $F_r \times F_s$ under R is clopen, by (2).

Consider now the case of an arbitrary clopen subset H of the product space $U \times U$. Because $\mathfrak{U}$ is a Boolean space, every clopen subset of $U \times U$ can be written as the union of finitely many products of clopen subsets of $\mathfrak{U}$. There is consequently a finite set I of pairs of elements from $\mathfrak{A}$ such that

$$H = \bigcup \{F_r \times F_s : (r, s) \in I\}. \quad (3)$$

It follows from (3) and the definition of the image of a set under the relation R that

$$R^*(H) = \{Z : R(X, Y, Z) \text{ for some } (X, Y) \in H\}$$
$$= \{Z : R(X, Y, Z) \text{ for some } (r, s) \in I \text{ and } (X, Y) \in F_r \times F_s\}$$
$$= \bigcup \{R^*(F_r \times F_s) : (r, s) \in I\}.$$

It was shown in the preceding paragraph that the sets $R^*(F_r \times F_s)$ are clopen in $\mathfrak{U}$. Since a union of finitely many clopen sets is clopen, it follows that $R^*(H)$ is clopen in $\mathfrak{U}$. $\square$

In order to prove that the fundamental relations in $\mathfrak{U}$ of rank at least two are continuous, it is helpful to prove first that the images of singleton points under these relations are closed sets.

Lemma 2.6. *For every fundamental relation R of rank $n + 1$ in the relational structure $\mathfrak{U}$ correlated with the given Boolean algebra with operators $\mathfrak{A}$, and for every sequence of points $X_0, \ldots, X_{n-1}$ in $\mathfrak{U}$, the image set*

$$R^*(X_0, \ldots, X_{n-1}) = \{Z \in U : R(X_0, \ldots, X_{n-1}, Z)\}$$

is a closed, and hence compact, subset of $\mathfrak{U}$.

Proof. Focus on the case of a ternary relation R in $\mathfrak{U}$ that is defined in terms of a binary operator $\circ$ in $\mathfrak{A}$. Let X and Y be points in $\mathfrak{U}$, and write

$$H = R^*(X, Y) = \{Z \in U : R(X, Y, Z)\}. \tag{1}$$

It is to be shown that H is a closed, compact subset of $\mathfrak{U}$. If a point W in $\mathfrak{U}$ does not belongs to H—that is to say, if $R(X, Y, W)$ fails to hold—then the complex product $X \circ Y$ cannot be included in W, by the definition of the relation R in $\mathfrak{U}$. Consequently, there must be elements r_W in X and s_W in Y such that the product $r_W \circ s_W$ (formed in $\mathfrak{A}$) does not belong to W, by the definition of the complex product $X \circ Y$. The set $F_{r_W \circ s_W}$ of all ultrafilters that contain the product $r_W \circ s_W$ is a clopen subset of $\mathfrak{U}$, by the definition of the topology on $\mathfrak{U}$, and this clopen set does not contain W because $r_W \circ s_W$ is not in W. The intersection of a system of clopen sets is closed in the topology on U, so the set

$$F = \bigcap \{F_{r_W \circ s_W} : W \in U \sim H\}$$

is a closed subset of $\mathfrak{U}$ that does not contain W for any point W in $U \sim H$. In other words, F is a closed set that is disjoint from $U \sim H$.

Consider now any point Z in H. Since $R(X, Y, Z)$ holds, by (1), the definition of the relation R in $\mathfrak{U}$ implies that the complex product $X \circ Y$ is included in Z. Consequently, the element $r_W \circ s_W$ must belong to Z for every point W in $U \sim H$, because r_W is in X, and s_W in Y. It follows that the ultrafilter Z belongs to the clopen set $F_{r_W \circ s_W}$ for every W in $U \sim H$, and therefore Z belongs to the intersection F of these clopen sets, by the definition of F. Thus, the set H is included in F. Combine this observation with those of the preceding paragraph to conclude that the set H coincides with F and is therefore closed.

Every closed subset of a compact topological space is compact, so the set H must be compact. $\square$

Lemma 2.7. *Every fundamental relation of rank at least two in the relational structure $\mathfrak{U}$ correlated with the given Boolean algebra with operators $\mathfrak{A}$ is continuous.*

Proof. Focus on the case of a ternary relation R in $\mathfrak{U}$ that is defined in terms of a binary operator $\circ$ in $\mathfrak{A}$. The proof requires a preliminary observation: for every element t in $\mathfrak{A}$, a pair (X, Y) of points from $\mathfrak{U}$ belongs to the inverse image $R^{-1}(F_t)$ of the clopen set F_t if and only if there are elements r in X and s in Y such that $r \circ s \leq t$ (in $\mathfrak{A}$). One direction of the argument is straightforward. Suppose such elements r and s exist. If $R(X, Y, Z)$ holds, then the complex product $X \circ Y$ is included in Z, by the definition of the relation R, and therefore the product $r \circ s$ belongs to Z, by the definition of the complex product and the assumption that r and s are in X and Y respectively. Since $r \circ s$ is below t, and Z is an ultrafilter, the element t must also belong to Z, and therefore Z must belong to the clopen set F_t. This is true for every Z such that $R(X, Y, Z)$ holds, so the entire image set

$$R^*(X, Y) = \{Z \in U : R(X, Y, Z)\} \tag{1}$$

is included in F_t. Consequently, the pair (X, Y) belongs to the inverse image $R^{-1}(F_t)$, by the definition of this inverse image.

To establish the reverse implication, suppose that no such elements r and s exist. We proceed to show that in this case the set

$$X \circ Y \cup \{-t\} = \{r \circ s : r \in X \text{ and } s \in Y\} \cup \{-t\} \tag{2}$$

has the finite meet property. Assume, accordingly, that

$$(r_i : i \in I) \quad \text{and} \quad (s_i : i \in I)$$

are finite systems of elements in X and Y respectively, write

$$r = \prod_i r_i \quad \text{and} \quad s = \prod_i s_i,$$

and observe that r belongs to X and s to Y, because X and Y are ultrafilters and are therefore closed under finite Boolean products. The element $r \circ s$ is not below t, by assumption, so

$$-t \cdot (r \circ s) \neq 0.$$

Since $r \leq r_i$ and $s \leq s_i$ for each i, the monotony of the operator $\circ$ implies that $r \circ s \leq r_i \circ s_i$ for each i, and therefore $r \circ s \leq \prod_i (r_i \circ s_i)$. It follows that

$$-t \cdot \prod_i (r_i \circ s_i) \neq 0,$$

as required.

Every subset of $\mathfrak{A}$ with the finite meet property can be extended to an ultrafilter, so the set in (2) can be extended to an ultrafilter Z in $\mathfrak{A}$, by the observations of the preceding paragraph. The complex product $X \circ Y$ is included in Z, by (2), so $R(X, Y, Z)$ holds, by the definition of R. On the other hand, the complement $-t$ is in Z, again by (2), and Z is an ultrafilter, so t does not belong to Z, and therefore Z does not belong to F_t. Apply the definition of the inverse image to conclude that the pair (X, Y) cannot belong to the inverse image $R^{-1}(F_t)$. This completes the proof of the preliminary observation.

In order to prove that R is continuous, it must be shown that for every open subset H of U, the inverse image set

$$R^{-1}(H) = \{(X, Y) : R(X, Y, Z) \text{ implies } Z \in H\}$$

is open in the product topology on $U \times U$. Consider first the case when H is a clopen subset of U, say $H = F_t$. If a pair (X, Y) belongs to the inverse image $R^{-1}(F_t)$, then there must be elements r in X and s in Y such that $r \circ s \leq t$, by the preliminary observation proved above. The ultrafilters X and Y belong to the clopen sets F_r and F_s respectively, by the definition of these clopen sets, so the pair (X, Y) belongs to the product clopen set $F_r \times F_s$. Furthermore, this product clopen set is included in the inverse image $R^{-1}(F_t)$. Indeed, it is not difficult to check that

$$F_r \times F_s \subseteq R^{-1}(F_r \circ F_s) \subseteq R^{-1}(F_t). \tag{3}$$

For the second inclusion, observe that the inequality $r \circ s \leq t$ implies that the product $F_r \circ F_s$ (formed in $\mathfrak{Cm}(U)$) is included in F_t, by Lemma 2.4, and therefore the inverse image under R of the product $F_r \circ F_s$ is included in the inverse image under R of F_t, Lemma 2.1. As regards the first inclusion, suppose X_0 is in F_r and Y_0 in F_s. The product $\{X_0\} \circ \{Y_0\}$ in $\mathfrak{Cm}(U)$ of the atoms $\{X_0\}$ and $\{Y_0\}$ is included in the product $F_r \circ F_s$, by the monotony of the operator $\circ$. In other words,

$$Z \in \{X_0\} \circ \{Y_0\} \qquad \text{implies} \qquad Z \in F_r \circ F_s,$$

or put somewhat differently,

$$R^*(X_0, Y_0) \subseteq F_r \circ F_s.$$

This last inclusion is precisely the condition required for (X_0, Y_0) to belong to the inverse image $R^{-1}(F_r \circ F_s)$.

It has been shown that every pair (X, Y) in $R^{-1}(F_t)$ belongs to a clopen subset $F_r \times F_s$ of $U \times U$ that in turn is included in $R^{-1}(F_t)$. Consequently, the inverse image set $R^{-1}(F_t)$ is a union of clopen sets and is therefore open in $U \times U$.

Consider now the general case when H is an arbitrary open subset of $\mathfrak{U}$. The space $\mathfrak{U}$ is Boolean, so H must be the union of a system

$$(F_t : t \in I) \tag{4}$$

of clopen sets. For each pair (X, Y) in the inverse image $R^{-1}(H)$, the image set in (1) is included in H, by the definition of the inverse image set. Consequently, this image set is included in the union of the system (4). The image set in (1) is compact, by Lemma 2.6, so there must be a finite subsystem of (4) whose union covers $R^*(X, Y)$. The union of finitely many clopen sets is clopen, so there is a clopen set F_t such that

$$R^*(X, Y) \subseteq F_t \subseteq H. \tag{5}$$

The first inclusion in (5) implies that the pair (X, Y) belongs to the inverse image $R^{-1}(F_t)$, by the definition of this inverse image. It was shown in the previous paragraph that this inverse image is an open subset of $\mathfrak{U}$. The second inclusion in (5) implies that $R^{-1}(F_t)$ is included in $R^{-1}(H)$, by Lemma 2.1. Conclusion: every pair (X, Y) in the inverse image $R^{-1}(H)$ belongs to an open set $R^{-1}(F_t)$ that is included in $R^{-1}(H)$, so $R^{-1}(H)$ is a union of open sets and is therefore open. Consequently, R is continuous. $\square$

The following theorem contains a summary of what has been accomplished in the preceding discussion and lemmas.

Theorem 2.8. *The relational structure correlated with a Boolean algebra with operators $\mathfrak{A}$ is a relational space under the topology induced on the relational structure by $\mathfrak{A}$.*

The relational space correlated with $\mathfrak{A}$ is called the (*first*) *dual*, or the *dual* (*relational*) *space*, of $\mathfrak{A}$. As was mentioned earlier, if $\mathfrak{U}$ is the dual of $\mathfrak{A}$, then we may employ topological terminology in speaking about $\mathfrak{U}$, referring for example to the *points* in $\mathfrak{U}$ (the elements in the universe of $\mathfrak{U}$) and the *clopen sets* in $\mathfrak{U}$ (the clopen sets in the topology on the universe of $\mathfrak{U}$).

We now change footing: starting with an arbitrary relational space $\mathfrak{U}$, we construct a Boolean algebra with operators $\mathfrak{A}$ in terms of $\mathfrak{U}$.

Lemma 2.9. *The set of clopen subsets of a relational space $\mathfrak{U}$ is a subuniverse of the complex algebra $\mathfrak{Cm}(U)$.*

Proof. Let A be the set of clopen subsets of $\mathfrak{U}$. Clearly, A is a subset of the universe of $\mathfrak{Cm}(U)$. The union of two clopen sets is clopen, and the complement of a clopen set is clopen, so A is closed under the Boolean operations of $\mathfrak{Cm}(U)$, namely union and complement. To see that A is closed under the operators of $\mathfrak{Cm}(U)$, focus on the case of a binary operator $\circ$ that is defined in terms of a ternary relation R in $\mathfrak{U}$. For two clopen subsets F and G in A, the product $F \circ G$ in $\mathfrak{Cm}(U)$ is defined to be the set

$$F \circ G = \{t \in U : R(r, s, t) \text{ for some } r \in F \text{ and } s \in G\}.$$

This is precisely the definition of the image set $R^*(F \times G)$, so

$$F \circ G = R^*(F \times G). \tag{1}$$

The definition of a relational space implies that the relation R is clopen; this means—since F and G are assumed to be clopen subsets of U— that the image set $R^*(F \times G)$ is a clopen subset of U. Thus, $F \circ G$ is a clopen set, by (1), so it belongs to A. It follows that A is closed under the operator $\circ$. $\square$

Observe that the assumed continuity of the relations in $\mathfrak{U}$ of rank at least two is not used in the preceding proof.

The complex algebra $\mathfrak{Cm}(U)$ is a Boolean algebra with operators, and this property is preserved under the passage to subalgebras. In particular, the subalgebra of $\mathfrak{Cm}(U)$ that has as its universe the set of the clopen subsets of $\mathfrak{U}$ is a Boolean algebra with operators. It is called the *(first) dual*, or the *dual algebra*, of $\mathfrak{U}$. Start with a Boolean algebra with operators $\mathfrak{A}$, and form the dual space $\mathfrak{U}$ consisting of ultrafilters in $\mathfrak{A}$. The dual algebra of $\mathfrak{U}$ consisting of the clopen subsets of $\mathfrak{U}$ is called the *second dual* of $\mathfrak{A}$.

Theorem 2.10. *The second dual of every Boolean algebra with operators $\mathfrak{A}$ is isomorphic to $\mathfrak{A}$. More explicitly, if $\mathfrak{U}$ is the dual space of $\mathfrak{A}$, and $\mathfrak{B}$ the dual algebra of $\mathfrak{U}$, and if*

$$\varphi(r) = F_r = \{X \in U : r \in X\}$$

for each r in $\mathfrak{A}$, then φ is an isomorphism from $\mathfrak{A}$ to $\mathfrak{B}$.

Proof. Essentially, this theorem has already been proved. We summarize the main points of the argument. The relational space $\mathfrak{U}$ that is the dual of $\mathfrak{A}$ has as its universe the set of ultrafilters in $\mathfrak{A}$. Each relation R of rank $n + 1$ in $\mathfrak{U}$ is correlated with an operator f of rank n in $\mathfrak{A}$ and is defined by the formula

$$R(X_0, \ldots, X_{n-1}, Z) \qquad \text{if and only if} \qquad f(X_0, \ldots, X_{n-1}) \subseteq Z$$

for every sequence of ultrafilters $X_0, \ldots, X_{n-1}, Z$ (where the occurrence of f on the right side of this equivalence denotes the extension of the operator f in $\mathfrak{A}$ to subsets of $\mathfrak{A}$). The topology on $\mathfrak{U}$ has as a base the set of all sets of the form F_r for r in $\mathfrak{A}$, and these sets prove to be precisely the clopen sets of the topology. The fundamental relations in $\mathfrak{U}$ are continuous (Lemma 2.7) and clopen (Lemma 2.5) under this topology.

The Boolean algebra with operators $\mathfrak{B}$ that is the dual of $\mathfrak{U}$ has as its universe the set of clopen subsets of $\mathfrak{U}$, that is to say, the universe consists of the sets of the form F_r for r in $\mathfrak{A}$. The Boolean operations are the set-theoretic ones of union and complement, while each operator f of rank n in $\mathfrak{B}$ is defined on a sequence of clopen sets $F_{r_0}, \ldots, F_{r_{n-1}}$ by

$$f(F_{r_0}, \ldots, F_{r_{n-1}}) = R^*(F_{r_0} \times \cdots \times F_{r_{n-1}})$$

(see the proof of Lemma 2.9), and the image set on the right side of this equation coincides with the set F_t when $t = f(r_0, \ldots, r_{n-1})$ in $\mathfrak{A}$ (see the proof of Lemma 2.5), so

$$f(F_{r_0}, \ldots, F_{r_{n-1}}) = F_t \qquad \text{if and only if} \qquad t = f(r_0, \ldots, r_{n-1}).$$

Consequently, $\mathfrak{B}$ is precisely the subalgebra of $\mathfrak{Cm}(U)$ that is the image of $\mathfrak{A}$ under the canonical embedding that maps each element r in $\mathfrak{A}$ to the element F_r in $\mathfrak{B}$. In particular, the canonical embedding is an isomorphism from $\mathfrak{A}$ to its second dual $\mathfrak{B}$. $\square$

Theorem 2.10 has a topological analogue. In order to formulate it, we need some terminology. Start with a relational space $\mathfrak{U}$, and form its dual algebra $\mathfrak{A}$ consisting of the clopen subsets of $\mathfrak{U}$. The dual space of $\mathfrak{A}$, consisting of the ultrafilters in $\mathfrak{A}$ and endowed with the topology induced by $\mathfrak{A}$, is called the *second dual of* $\mathfrak{U}$. We shall prove that there is a bijection ϑ from the universe of $\mathfrak{U}$ to the universe of the second dual of $\mathfrak{U}$ that preserves the algebraic and topological structure of $\mathfrak{U}$ in the sense that ϑ is a relational structure isomorphism and a topological homeomorphism. We shall call such a function a *homeo-isomorphism*, and we shall say that two relational spaces are *homeo-isomorphic* if there exists a homeo-isomorphism mapping one of them onto the other.

Theorem 2.11. *The second dual of every relational space* $\mathfrak{U}$ *is homeo-isomorphic to* $\mathfrak{U}$. *More explicitly, if* $\mathfrak{A}$ *is the dual algebra of* $\mathfrak{U}$, *and* $\mathfrak{V}$ *the dual space of* $\mathfrak{A}$, *and if*

$$\vartheta(r) = \{F \in A : r \in F\}$$

for each r *in* $\mathfrak{U}$, *then* ϑ *is a homeo-isomorphism from* $\mathfrak{U}$ *to* $\mathfrak{V}$.

Proof. The universe of the dual algebra $\mathfrak{A}$ of $\mathfrak{U}$ is defined to be the set of all clopen subsets of $\mathfrak{U}$, and the universe of the dual space $\mathfrak{V}$ of $\mathfrak{A}$ is defined to be the set of all ultrafilters in $\mathfrak{A}$. The set

$$X_r = \{F \in A : r \in F\} \tag{1}$$

is easily seen to be such an ultrafilter in $\mathfrak{A}$, so the function ϑ defined by

$$\vartheta(r) = X_r \tag{2}$$

maps the universe of $\mathfrak{U}$ into the universe of $\mathfrak{V}$.

It is not difficult to see that ϑ is one-to-one. If r and s are distinct points in $\mathfrak{U}$, then there is a clopen set F in $\mathfrak{U}$ that contains r but not s, because $\mathfrak{U}$ has the topological structure of a Boolean space and therefore the clopen sets separate points. The set F—which is an element in $\mathfrak{A}$—belongs to the ultrafilter X_r, but not to the ultrafilter X_s, so

$$\vartheta(r) = X_r \neq X_s = \vartheta(s).$$

To check that ϑ is onto, consider a point Y in the second dual $\mathfrak{V}$. This point must be an ultrafilter in $\mathfrak{A}$, by the definition of $\mathfrak{V}$. In other words, Y must be a maximal class of non-empty clopen subsets of $\mathfrak{U}$ that is upward closed and closed under finite intersections. The compactness of the topology on $\mathfrak{U}$ implies that every non-empty system of closed sets with the finite intersection property has a non-empty intersection. In particular, since the ultrafilter Y has the finite intersection property, the intersection of the system of all sets in Y is non-empty. Let r be any point in this intersection. The point r belongs to every (clopen) set in Y, so every set in Y must belong to the ultrafilter X_r, by (1). Thus, Y is included in X_r. Since Y is assumed to be an ultrafilter, it follows that $Y = X_r$, by the maximality of Y, and therefore

$$\vartheta(r) = X_r = Y.$$

Thus, ϑ is onto.

A bijection from a Boolean space to a compact Hausdorff is a homeomorphism if and only if it maps clopen sets to clopen sets (see Corollary 3 on p. 316 of [10]). Consequently, to prove that ϑ is a homeomorphism, it suffices to show that the image under ϑ of each clopen set in $\mathfrak{U}$ is a clopen set in $\mathfrak{V}$. The image under the mapping ϑ of a clopen set F in $\mathfrak{U}$ is the set

$$\{X_r : r \in F\}, \tag{3}$$

by the definition of ϑ in (2). To say that r belongs to F is equivalent to saying that F belongs to X_r, by (1), so (3) may be rewritten as

$$\{X_r : r \in U \text{ and } F \in X_r\}. \tag{4}$$

The elements in $\mathfrak{V}$ are precisely the sets of the form X_r for r in $\mathfrak{U}$, by the conclusion of the previous paragraph, so (4) may in turn be rewritten as

$$\{Y \in V : F \in Y\}. \tag{5}$$

Because $\mathfrak{V}$ is assumed to be the dual space of the algebra $\mathfrak{A}$, each clopen subset of $\mathfrak{V}$ is determined by an element F in $\mathfrak{A}$ in the sense that it can be written as a set of the form (5) (see the remarks preceding Lemma 2.5). Thus, the set in (5) is clopen in the topology of $\mathfrak{V}$. Combine these observations to conclude that the image under ϑ of the clopen set F in $\mathfrak{U}$ is the clopen set (5) in $\mathfrak{V}$, so the mapping ϑ is clopen and therefore a homeomorphism, as claimed.

It remains to prove that the mapping ϑ is a relational structure isomorphism in the sense that it isomorphically preserves the fundamental relations of the relational structures. Focus on the case of a ternary relation R. In view of (2), it must be shown that the equivalence

$$R(r, s, t) \qquad \text{if and only if} \qquad R(X_r, X_s, X_t) \tag{6}$$

holds for all elements r, s, and t in $\mathfrak{U}$. (The first occurrence of R in (6) refers to the relation in $\mathfrak{U}$, while the second occurrence refers to the relation in $\mathfrak{V}$.) The assertion

$$R(X_r, X_s, X_t) \tag{7}$$

means, by the definition of $\mathfrak{V}$ as the dual space of $\mathfrak{A}$, that the complex product

$$X_r \circ X_s = \{F \circ G : F \in X_r \text{ and } G \in X_s\}$$

is included in the set X_t, that is to say,

$$F \in X_r \quad \text{and} \quad G \in X_s \quad \text{implies} \quad F \circ G \in X_t$$

for all elements F and G in $\mathfrak{A}$. This implication may be rewritten in the form

$$r \in F \quad \text{and} \quad s \in G \quad \text{implies} \quad t \in F \circ G \tag{8}$$

for all clopen subsets F and G of $\mathfrak{U}$. Since $\mathfrak{A}$ is the dual algebra of $\mathfrak{U}$, the product $F \circ G$ in $\mathfrak{A}$ is defined to be the image of the pair of clopen sets (F, G) under the relation R in $\mathfrak{U}$,

$$\begin{aligned} F \circ G &= R^*(F \times G) \\ &= \{w \in U : R(p, q, w) \text{ for some } p \in F \text{ and } q \in G\}, \end{aligned}$$

so the implication in (8) is equivalent to the implication

$$r \in F \text{ and } s \in G \text{ implies } R(p, q, t) \text{ for some } p \in F \text{ and } q \in G \tag{9}$$

for all clopen subsets F and G of $\mathfrak{U}$. These observations show that the validity of (7) is equivalent to the validity of (9) for all clopen sets F and G in $\mathfrak{U}$.

The implication from left to right in (6) is now easy to establish. If

$$R(r, s, t) \tag{10}$$

holds, then for all clopen sets F and G in $\mathfrak{U}$ containing the points r and s respectively, there are always points p in F and q in G such that $R(p, q, t)$ holds, namely the points $p = r$ and $q = s$, by (10). Consequently, (9) is valid for all clopen sets F and G in $\mathfrak{U}$, and therefore (7) holds.

In order to establish the reverse implication in (6), assume that (10) fails, with the intention of showing that (7) fails. For every point w in $\mathfrak{U}$, the validity of $R(r, s, w)$ implies that w must be different from t, by the assumption that (10) fails. The clopen subsets of $\mathfrak{U}$ separate points, so for such a point w there must be a clopen set H_w in $\mathfrak{U}$ that contains w, but does not contain t. The union of these clopen sets is an open set

$$H = \bigcup \{ H_w : w \in U \text{ and } R(r, s, w) \}$$

in $\mathfrak{U}$ that does not contain t. The inverse image of H under the relation R in $\mathfrak{U}$, that is to say, the set

$$R^{-1}(H) = \{ (p, q) \in U \times U : R(p, q, w) \text{ implies } w \in H \}, \qquad (11)$$

is open in the product topology on $U \times U$, by the assumed continuity of R. Moreover, the pair (r, s) belongs to this inverse image. Indeed, if w is any point in $\mathfrak{U}$ such that $R(r, s, w)$, then w belongs to the clopen set H_w, by the definition of this set, and therefore w belongs to H, by the definition of H; consequently, (r, s) is in $R^{-1}(H)$, by (11). The products of pairs of clopen subsets of $\mathfrak{U}$ form a base for the product topology on $U \times U$. Since $R^{-1}(H)$ is open in this topology and contains the point (r, s), there must be clopen sets F and G in $\mathfrak{U}$ for which

$$r \in F, \qquad s \in G, \qquad \text{and} \qquad F \times G \subseteq R^{-1}(H). \qquad (12)$$

The clopen sets F and G satisfy the hypothesis of (9), by the first part of (12), but not the conclusion of (9), by the last part of (12). Indeed, if we had $R(p, q, t)$ for some p in F and q in G, then the pair (p, q) would belong to $F \times G$, and therefore also to $R^{-1}(H)$, by the final part of (12); this would force the point t to be in H, by (11), in contradiction to the fact that t does not belong to H. Conclusion: if (10) fails, then there are clopen sets F and G for which (9) fails, and therefore (7) fails, as claimed. Thus, ϑ isomorphically preserves the ternary relation R.

The proof that the mapping ϑ isomorphically preserves a unary relation R is somewhat different in character than the preceding argument. It is to be shown that the equivalence

$$R(t) \qquad \text{if and only if} \qquad R(X_t) \qquad (13)$$

holds for all elements t in $\mathfrak{U}$, by the definition of ϑ in (2). (The first occurrence of R in (13) refers to the relation in $\mathfrak{U}$, while the second occurrence refers to the relation in $\mathfrak{V}$.) The point X_t belongs to the relation R in $\mathfrak{V}$ just in case the corresponding distinguished constant in $\mathfrak{A}$ belongs to X_t, by the definition of the relational space $\mathfrak{V}$ correlated with the algebra $\mathfrak{A}$. The corresponding distinguished constant in $\mathfrak{A}$ is just the relation R from $\mathfrak{U}$, by the definition of the complex algebra $\mathfrak{Cm}(U)$ and the fact that $\mathfrak{A}$ is a subalgebra of the complex algebra; and R belongs to X_t just in case t belongs to R, by (1). Summarizing, we have

$$X_t \in R \quad \text{if and only if} \quad R \in X_t,$$
$$\text{if and only if} \quad t \in R$$

(where the first occurrence of R refers to the unary relation, or set, in $\mathfrak{V}$, and the second and third to the unary relation, or set, in $\mathfrak{U}$). The equivalence of the first formula with the last is just another way of expressing (13). This completes the proof that the function ϑ is a homeo-isomorphism from the relational space $\mathfrak{U}$ to its second dual $\mathfrak{V}$. $\square$

There is another version of the notion of a relational space that proves to be equivalent to the version given in Definition 2.2, namely (a slightly modified form of) the definition that is used in Goldblatt [13], pp. 184–185, restricted to Boolean algebras with operators. (An earlier version of Goldblatt's approach, for Boolean algebras with a single unary operator, is given in Section 10 of Goldblatt [12], and it is remarked that his construction is an adaptation of one given in Section III of Lemmon [24].)

Definition 2.12. A *weak relational space* is a relational structure $\mathfrak{U}$, together with a topology on the universe U, such that U is a Boolean space under the topology, the relations in $\mathfrak{U}$ are clopen, and the relations R in $\mathfrak{U}$ of rank $n + 1$ (with $n \geq 1$) are *weakly continuous* in the sense that the sets

$$\{(r_0, \ldots, r_{n-1}) \in U^n : R(r_0, \ldots, r_{n-1}, t)\}$$

are closed (in the product space U^n) for every element t in $\mathfrak{U}$. $\square$

It turns out that, on the basis of the remaining conditions in the two definitions, a relation in $\mathfrak{U}$ of rank at least two is continuous if and only if it is weakly continuous. One direction of this implication is not difficult to prove.

Theorem 2.13. *Every relational space $\mathfrak{U}$ is a weak relational space. In particular, every relation in $\mathfrak{U}$ of rank at least two is weakly continuous.*

Proof. Focus on the case of a ternary relation R in $\mathfrak{U}$ that is, by assumption, continuous. To show that R is weakly continuous, fix an element t in $\mathfrak{U}$. The singleton set $\{t\}$ is closed in the topology on $\mathfrak{U}$, because $\mathfrak{U}$ is a Hausdorff space and singleton subsets are always closed in a Hausdorff space. Consequently, the complement $\sim\{t\}$ is open. The inverse image under R of this complement, the set

$$R^{-1}(\sim\{t\}) = \{(r,s) \in U \times U : R(r,s,\bar{t}) \text{ implies } \bar{t} \in \sim\{t\}\} \quad (1)$$
$$= \{(r,s) \in U \times U : R(r,s,\bar{t}) \text{ implies } \bar{t} \neq t\},$$

is therefore open in the product topology on $U \times U$, because R is continuous.

It must be shown that the set

$$H_t = \{(r,s) \in U \times U : R(r,s,t)\} \quad (2)$$

is closed in U, or what amounts to the same thing, the complement $\sim H_t$ is open. This is accomplished by establishing the equality

$$\sim H_t = R^{-1}(\sim\{t\}). \quad (3)$$

If a pair (r,s) belongs to the complement $\sim H_t$, then $R(r,s,t)$ must fail in $\mathfrak{U}$, by (2), and therefore $R(r,s,\bar{t})$ always implies that $\bar{t} \neq t$. Thus, the pair (r,s) belongs to the inverse image $R^{-1}(\sim\{t\})$, by (1), so the left side of (3) is included in the right side. As regards the reverse inclusion, if a pair (r,s) belongs to the inverse image $R^{-1}(\{\sim\{t\})$, then $R(r,s,\bar{t})$ always implies that $\bar{t} \neq t$, by (1), and therefore $R(r,s,t)$ cannot hold. It follows that the pair (r,s) cannot belong to H_t, by (2), so the pair must belong to the complement $\sim H_t$. This establishes (3). Since the set $R^{-1}(\sim\{t\})$ is open, by the remarks above, the set H_t must closed, by (3). This is true for every element t in $\mathfrak{U}$, so the relation R is weakly continuous. $\square$

The proof of the converse assertion that every weak relational space is a relational space appears to be significantly more involved. We shall give an indirect proof that depends on several of the main results of this section. First of all, Lemma 2.9 remains true for every weak relational space $\mathfrak{U}$. Indeed, as has already been noted, the proof of the lemma does not make use of the assumption that the fundamental relations of rank at least two are continuous (or even weakly continuous). The subalgebra of $\mathfrak{Cm}(U)$ that has as its universe the set of clopen subsets of $\mathfrak{U}$ is therefore a Boolean algebra with operators; it is called the (*first*) *dual* of the weak relational space $\mathfrak{U}$. Write $\mathfrak{A}$ for this dual.

The algebra $\mathfrak{A}$ has a dual relational space $\mathfrak{V}$ (and not just a dual *weak* relational space), by Theorem 2.8. In particular, the fundamental relations of rank at least two in $\mathfrak{V}$ are continuous, by Lemma 2.7 applied to $\mathfrak{V}$. Call $\mathfrak{V}$ the *second dual* of the weak relational space $\mathfrak{U}$. The key step in the argument is to show that $\mathfrak{U}$ is homeo-isomorphic to $\mathfrak{V}$, that is to say, Theorem 2.11 continues to hold for weak relational spaces. The proof that the function ϑ defined in the statement of that theorem is a homeomorphism from the topological space U to the topological space V, and the proof that for a unary relation R,

$$R(t) \qquad \text{if and only if} \qquad R(X_t)$$

remain unchanged. Indeed, the assumption that the fundamental relations of rank at least two are continuous or weakly continuous is not used in those parts of the argument.

The proof that ϑ isomorphically preserves every fundamental relation R of rank $n + 1 \geq 2$, and is therefore a homeo-isomorphism, is of course different from the earlier proof, since that proof uses the assumption that R is continuous. The proof below uses only the assumption that R is weakly continuous. (The argument is essentially the one given in Goldblatt [13], starting at the middle of p. 191.)

Consider the case of a ternary relation R. The equivalence

$$R(r, s, t) \qquad \text{if and only if} \qquad R(X_r, X_s, X_t) \qquad (6)$$

must be established. As in the proof of Theorem 2.11, the right side of (6) holds in $\mathfrak{V}$ just in case the implication

$$r \in F \quad \text{and} \quad s \in G \qquad \text{implies} \qquad t \in F \circ G \qquad (8)$$

is valid for all clopen subsets F and G of $\mathfrak{U}$, where

$$F \circ G = R^*(F \times G)$$
$$= \{w \in U : R(p, q, w) \text{ for some } p \in F \text{ and } q \in G\}.$$

Assume first that the left side of (6) is true in $\mathfrak{U}$, and consider arbitrary clopen subsets F and G of $\mathfrak{U}$. If r is in F and s in G, then since $R(r, s, t)$ holds, the element t must belong to the set $F \circ G$, by the definition of this set, and therefore the implication in (8) holds. Thus, the right side of (6) is true in $\mathfrak{V}$.

Assume next that the left side of (6) is false in $\mathfrak{U}$. In this case, the pair (r, s) does not belong to the set

$$H_t = \{(p, q) \in U \times U : R(p, q, t)\},$$

so (r, s) must belong to the complement $\sim H_t$ (in $U \times U$). The set H_t is closed, by the assumed weak continuity of the relation R, so the complement $\sim H_t$ is open. The products of clopen subsets of $\mathfrak{U}$ form a base for the product topology on $U \times U$, so there must be clopen subsets F and G of $\mathfrak{U}$ such that

$$r \in F, \qquad s \in G, \qquad \text{and} \qquad F \times G \subseteq \sim H_t.$$

The pair of sets F and G therefore satisfies the hypothesis of (8), but not the conclusion. Indeed, if p and q are any elements in F and G respectively, then the pair (p, q) belongs to the set $\sim H_t$, and therefore $R(p, q, t)$ is false, by the definition of H_t; consequently, t cannot belong to the set $F \circ G$, by the definition of this set. The failure of (8) for the particular clopen sets F and G implies that the right side of (6) is false in $\mathfrak{V}$.

It has been shown that the function ϑ is a homeo-isomorphism from $\mathfrak{U}$ to $\mathfrak{V}$, so Theorem 2.11 continues to hold for weak relational spaces $\mathfrak{U}$ and their second duals $\mathfrak{V}$. As has already been pointed out, the second dual $\mathfrak{V}$ is a relational space, and not just a weak relational space. In particular, the fundamental relations of rank at least two in $\mathfrak{V}$ are continuous. Since $\mathfrak{U}$ is homeo-isomorphic to $\mathfrak{V}$, it follows that the fundamental relations of rank at least two in $\mathfrak{U}$ must also be continuous, so $\mathfrak{U}$ is in fact a relational space. The following theorem has been proved.

Theorem 2.14. *Every weak relational space $\mathfrak{U}$ is a relational space. In particular, the fundamental relations of rank at least two in $\mathfrak{U}$ are continuous.*

Goldblatt [13] shows for his notion of a relational space that every bounded distributive lattice with operators is isomorphic to its second dual, and every relational space is homeo-isomorphic to its second dual (see Theorems 2.2.3 and 2.2.4 in [13]). It follows from Theorems 2.13 and 2.14 above that for Boolean algebras with operators, the approach we have taken is equivalent to the approach that Goldblatt has taken.

There is one more topological observation concerning relational spaces that is worth making and that will be used later: the fundamental relations are closed subsets of the appropriate product space. For the unary relations, this follows directly from Definition 2.2, but for relations of rank at least two it requires proof.

Theorem 2.15. *If $\mathfrak{U}$ is a relational space, then every fundamental relation in $\mathfrak{U}$ of rank $n \geq 2$ is a closed subset of the product spaces U^n.*

Proof. Focus on the case when R is a ternary relation in $\mathfrak{U}$. In order to prove that R is a closed subset of $U \times U \times U$, it suffices to shown that the complement of R is open. This is accomplished by demonstrating that for every triple of elements (r, s, t) not in R, there are clopen subsets F, G, and H of $\mathfrak{U}$ such that

$$(r, s, t) \in F \times G \times H \qquad \text{and} \qquad F \times G \times H \subseteq \sim R. \qquad (1)$$

Assume that the triple (r, s, t) is not in R. The set

$$K = \{(u, v) \in U \times U : R(u, v, t)\} \qquad (2)$$

is closed in $U \times U$, by Theorem 2.13, so the complement of K is open; and this complement contains the pair (r, s), by assumption. The products of clopen subsets of $\mathfrak{U}$ form a base for the product topology on $U \times U$, so there must be clopen sets F and G in $\mathfrak{U}$ containing the points r and s respectively such that $F \times G$ is included in $\sim K$. The set

$$R^*(F \times G) = \{w \in U : R(u, v, w) \text{ for some } u \in F \text{ and } v \in G\} \qquad (3)$$

is clopen in $\mathfrak{U}$, by the assumption that $\mathfrak{U}$ is a relational space (see Definition 2.2). The complement

$$\begin{aligned}
H &= \sim R^*(F \times G) \\
&= \{w \in U : R(u, v, z) \text{ and } u \in F \text{ and } v \in G \text{ implies } z \neq w\} \qquad (4)
\end{aligned}$$

is therefore also clopen in $\mathfrak{U}$.

Observe that t belongs to H. For the proof, consider points u in F and v in G, and suppose that z is a point in $\mathfrak{U}$ such that $R(u, v, z)$ holds. Since $F \times G$ is included in $\sim K$, the pair (u, v) must belong to $\sim K$, and therefore the point z must be different from t, by (2). It follows

by (4) that t must be in H. This observation and the definitions of the
sets F and G imply that the triple (r, s, t) belongs to the product

$$F \times G \times H. \tag{5}$$

To check that (5) is included in $\sim R$, as is required in (1), consider
any triple (u, v, w) in (5). The point w belongs to the set H, by the
definition of the product (5), so w must belong to the complement of
the set $R^*(F \times G)$, by (4). On the other hand, the points u and v are
in F and G respectively, so $R(u, v, w)$ must fail to hold in $\mathfrak{U}$, by (3).
This shows that the triple (u, v, w) belongs to the complement of R,
which completes the proof of (1). The relation $\sim R$ is therefore an open
subset of $U \times U \times U$, so R is closed. $\square$

2.4 Duality for Homomorphisms

The duality between Boolean algebras with operators and relational
spaces carries with it a duality between structure preserving functions on the algebras and structure preserving functions on the spaces.
The structure preserving functions on the algebras are just homomorphisms (in the algebraic sense of the word), while the structure
preserving functions on the spaces are continuous bounded homomorphisms (see Definition 1.8). Recall that a mapping ϑ from a topological
space V to a topological space U is said to be *continuous* if the inverse
image

$$\vartheta^{-1}(F) = \{u \in V : \vartheta(u) \in F\}$$

of each open subset F of U is open in V. When the spaces in question
are Boolean, then it suffices to check that the inverse image of each
clopen set is clopen (see Lemma 1 on p. 313 of [10]).

   The first task is to show that continuous bounded homomorphisms
give rise to homomorphisms. The proof is similar in flavor to the proof
of Theorem 1.9. That theorem is purely algebraic in formulation and
proof, with no reference to topologies on relational structures. The theorem we want refers also to the topologies and says that if the given
bounded homomorphism ϑ is in fact a continuous mapping with respect to the topologies, then an appropriate restriction of the mapping
defined in Theorem 1.9 is a homomorphism from the dual algebra of $\mathfrak{U}$
(the subalgebra of $\mathfrak{Cm}(U)$ consisting of the clopen subsets of $\mathfrak{U}$) to the

dual algebra of $\mathfrak{V}$ (the subalgebra of $\mathfrak{Cm}(V)$ consisting of the clopen subsets of $\mathfrak{V}$).

Theorem 2.16. *Let $\mathfrak{U}$ and $\mathfrak{V}$ be relational spaces, and $\mathfrak{A}$ and $\mathfrak{B}$ their respective dual algebras. If ϑ is a continuous bounded homomorphism from $\mathfrak{V}$ into $\mathfrak{U}$, then the function φ defined on elements F in $\mathfrak{A}$ by*

$$\varphi(F) = \vartheta^{-1}(F) = \{u \in V : \vartheta(u) \in F\}$$

is a homomorphism from $\mathfrak{A}$ to $\mathfrak{B}$. Moreover, φ is one-to-one if and only if ϑ is onto, and φ is onto if and only if ϑ is one-to-one.

Proof. The function ψ defined on subsets X of U by

$$\psi(X) = \vartheta^{-1}(X) = \{u \in V : \vartheta(u) \in X\}$$

is a complete homomorphism from $\mathfrak{Cm}(U)$ to $\mathfrak{Cm}(V)$, by Theorem 1.9. It is clear from this definition that the function φ defined in the statement of the theorem is the restriction of ψ to the set of all clopen subsets of $\mathfrak{U}$. Consequently, φ is a homomorphism from the subalgebra $\mathfrak{A}$ of $\mathfrak{Cm}(U)$ consisting of the clopen subsets of $\mathfrak{U}$ (see Lemma 2.9) into $\mathfrak{Cm}(V)$. The bounded homomorphism ϑ is assumed to be continuous, and the inverse image of a clopen set under a continuous function is again a clopen set (see pp. 312–313 of [10]), so the inverse image $\vartheta^{-1}(F)$ of a clopen subset F of $\mathfrak{U}$ is always a clopen subset of $\mathfrak{V}$. Therefore, φ actually maps $\mathfrak{A}$ into the subalgebra $\mathfrak{B}$ of $\mathfrak{Cm}(V)$ that consists of the clopen subsets of $\mathfrak{V}$.

Consider now the following statements: (1) ϑ is onto; (2) $U \sim \vartheta(V)$ is empty; (3) every clopen subset of $U \sim \vartheta(V)$ is empty; (4) if F is a clopen subset of U such that $\vartheta^{-1}(F)$ is empty, then F must be empty; (5) φ is one-to-one. Each of these statements is equivalent to its neighbor. Indeed, (1) is obviously equivalent to (2), and (2) obviously implies (3). To see that (3) implies (2), observe that $\vartheta(V)$ is the continuous image of the compact set V, so $\vartheta(V)$ is compact and therefore closed in U (see Lemma 3 on p. 314 and Lemma 1 on p. 272 of [10]). It follows that the difference $U \sim \vartheta(V)$ is open, and is therefore the union of a system of clopen sets. If each of these clopen sets is empty, as is asserted in (3), then clearly their union must also be empty, and therefore (2) holds. Statement (4) is really just a rephrasing of (3): to say of a subset F of U that $\vartheta^{-1}(F)$ is empty is to say that F does not contain any elements in the range of ϑ, or what amounts to the same thing, that F is a subset of $U \sim \vartheta(V)$. As regards the equivalence of (4) and (5), recall that a

Boolean homomorphism is one-to-one if and only if its kernel contains only the zero element. The kernel of φ is the set of clopen subsets F of U such that $\vartheta^{-1}(F)$ is empty, so this kernel contains only the zero element $\varnothing$ just in case (4) holds. The equivalence of (1) and (5) proves that φ is one-to-one if and only if ϑ is onto.

To prove the dual assertion, consider the following statements: (1) ϑ is one-to-one; (2) the inverse images under ϑ of the clopen subsets of U separate points in V (in the sense that for any two distinct points, the inverse image of some clopen subset of U contains one of the two points but not the other); (3) every clopen subset of V is the inverse image under ϑ of some clopen subset of U; (4) φ is onto. Each of these statements is equivalent to its neighbor. To establish the equivalence of (1) and (2), consider distinct points u and v in V. If (1) holds, then $\vartheta(u)$ and $\vartheta(v)$ are distinct points in U. The clopen subsets of U separate points (because U is a Boolean space), so there is a clopen subset F of U that contains $\vartheta(u)$ but not $\vartheta(v)$. The inverse image $\vartheta^{-1}(F)$ is a clopen subset of V (because ϑ is continuous), and it contains u but not v, so (2) holds. Conversely, if (2) holds, then there is a clopen subset F of U such that $\vartheta^{-1}(F)$ contains u but not v. Consequently, the set F contains $\vartheta(u)$ but not $\vartheta(v)$, so $\vartheta(u)$ and $\vartheta(v)$ must be distinct. Thus, (1) holds. To see that (2) implies (3), observe that the inverse images under ϑ of the clopen subsets of U constitute a Boolean algebra of clopen subsets of V under the operations of union and complement. In more detail, if F and G are clopen subsets of U, then so are $F \cup G$ and $U \sim F$; since

$$\vartheta^{-1}(F) \cup \vartheta^{-1}(G) = \vartheta(F \cup G) \quad \text{and} \quad \vartheta^{-1}(U \sim F) = V \sim \vartheta^{-1}(F),$$

it follows that the set of inverse images of clopen sets is closed under the operations of union and complement, and is therefore a Boolean algebra under these operations. This must be the Boolean algebra of all clopen subsets of V, by (2) and the fact that in a compact space, a Boolean algebra of clopen sets that separates points must be the Boolean algebra of all clopen subsets of the space (see Lemma 1 on p. 305 of [10]). The reverse implication from (3) to (2) follows from the fact that V is a Boolean space, and therefore the clopen subsets of V separate points. Finally, the equivalence of (3) and (4) follows from the definition of φ. The equivalence of (1) and (4) proves that ϑ is one-to-one if and only if φ is onto. $\square$

Goldblatt [13] shows (at the bottom of p. 193) that the function φ in the statement of Theorem 2.16 is a homomorphism from $\mathfrak{A}$ to $\mathfrak{B}$. He also points out that if ϑ is onto or one-to-one, then φ is one-to-one or onto respectively. (An earlier version of Goldblatt's result, for spaces with a single binary relation, is given in Theorem 5.9 of Goldblatt [12].)

It is also possible to formulate a version of Theorem 2.16 that refers not to relational spaces and their dual algebras, but rather to Boolean algebras with operators and their dual spaces.

Corollary 2.17. *Let $\mathfrak{A}$ and $\mathfrak{B}$ be Boolean algebras with operators, and $\mathfrak{U}$ and $\mathfrak{V}$ their respective dual spaces. If ϑ is a continuous bounded homomorphism from $\mathfrak{V}$ to $\mathfrak{U}$, then the function ψ defined on elements r in $\mathfrak{A}$ by*

$$\psi(r) = s \qquad \text{if and only if} \qquad \vartheta^{-1}(F_r) = G_s,$$

where

$$F_r = \{X \in U : r \in X\} \qquad \text{and} \qquad G_s = \{Y \in V : s \in Y\},$$

is a homomorphism from $\mathfrak{A}$ into $\mathfrak{B}$. Moreover, ψ is one-to-one if and only if ϑ is onto, and ψ is onto if and only if ϑ is one-to-one.

Proof. Let $\mathfrak{A}^*$ and $\mathfrak{B}^*$ be the dual algebras of the dual relational spaces $\mathfrak{U}$ and $\mathfrak{V}$ respectively. Thus, $\mathfrak{A}^*$ is the second dual of $\mathfrak{A}$, and $\mathfrak{B}^*$ is the second dual of $\mathfrak{B}$, by the definition of the second duals. The functions φ_1 and φ_2 defined by

$$\varphi_1(r) = F_r \qquad \text{and} \qquad \varphi_2(s) = G_s \tag{1}$$

for r in $\mathfrak{A}$ and s in $\mathfrak{B}$ are isomorphisms from $\mathfrak{A}$ to $\mathfrak{A}^*$ and from $\mathfrak{B}$ to $\mathfrak{B}^*$ respectively, by Theorem 2.10.

Assume that ϑ is a continuous bounded homomorphism from $\mathfrak{V}$ to $\mathfrak{U}$. The function φ defined by

$$\varphi(F) = \vartheta^{-1}(F) = \{Y \in V : \vartheta(Y) \in F\} \tag{2}$$

is a homomorphism from $\mathfrak{A}^*$ to $\mathfrak{B}^*$, by the first part of Theorem 2.16 (with $\mathfrak{A}^*$ and $\mathfrak{B}^*$ in place of $\mathfrak{A}$ and $\mathfrak{B}$, and Y in place of u). Moreover, φ is one-to-one or onto if and only if ϑ is onto or one-to-one respectively, by the second part of Theorem 2.16. The composition

$$\psi = \varphi_2^{-1} \circ \varphi \circ \varphi_1 \tag{3}$$

is therefore a homomorphism from $\mathfrak{A}$ to $\mathfrak{B}$ (see the diagram below). Moreover, since φ_1 and φ_2 are bijections, the homomorphism ψ is one-to-one or onto if and only the homomorphism φ is one-to-one or onto respectively. Consequently, ψ is one-to-one or onto if and only ϑ is onto or one-to-one respectively.

$$\mathfrak{A} \xrightarrow{\ \psi\ } \mathfrak{B}$$
$$\varphi_1 \downarrow \qquad\qquad \downarrow \varphi_2$$
$$\mathfrak{A}^* \xrightarrow{\ \varphi\ } \mathfrak{B}^*$$

If $\vartheta^{-1}(F_r) = G_s$, then $\varphi(F_r) = G_s$, by (2), and therefore

$$\psi(r) = (\varphi_2^{-1} \circ \varphi \circ \varphi_1)(r) = \varphi_2^{-1}(\varphi(\varphi_1(r)))$$
$$= \varphi_2^{-1}(\varphi(F_r)) = \varphi_2^{-1}(G_s) = s,$$

by (3) and (1). On the other hand, if $\psi(r) = s$, then

$$\varphi(F_r) = (\varphi_2 \circ \psi \circ \varphi_1^{-1})(F_r) = \varphi_2(\psi(\varphi_1^{-1}(F_r)))$$
$$= \varphi_2(\psi(r)) = \varphi_2(s) = G_s,$$

by (3) and (1), and therefore $\vartheta^{-1}(F_r) = G_s$, by (2). Thus, the homomorphism ψ is determined by the equivalence stated in the corollary. $\square$

The function φ in Theorem 2.16 is called the (*first*) *dual*, or the *dual homomorphism*, of the continuous bounded homomorphism ϑ. The equation defining this dual in the statement of the theorem can be reformulated as an equivalence, namely

$$u \in \varphi(F) \qquad \text{if and only if} \qquad u \in \vartheta^{-1}(F)$$

for all elements u in $\mathfrak{V}$ and F in $\mathfrak{A}$. This equivalence, in turn, may be written in the form

$$u \in \varphi(F) \qquad \text{if and only if} \qquad \vartheta(u) \in F$$

for all elements u in $\mathfrak{V}$ and F in $\mathfrak{A}$. Thus, the dual of ϑ is the function φ from $\mathfrak{A}$ to $\mathfrak{B}$ that is determined by the preceding equivalence.

Theorem 2.16 says that every continuous bounded homomorphism between relational spaces determines a dual homomorphism between the dual algebras of the spaces. The converse is also true: every homomorphism between the dual algebras induces a continuous bounded homomorphism between the relational spaces. We begin by proving the corresponding statement about Boolean algebras with operators and their dual spaces.

Theorem 2.18. *Let* $\mathfrak{A}$ *and* $\mathfrak{B}$ *be Boolean algebras with operators, and* $\mathfrak{U}$ *and* $\mathfrak{V}$ *their respective dual spaces. If* φ *is a homomorphism from* $\mathfrak{A}$ *into* $\mathfrak{B}$, *then the function* ϑ *defined on elements* Y *in* $\mathfrak{V}$ *by*

$$\vartheta(Y) = \varphi^{-1}(Y) = \{r \in A : \varphi(r) \in Y\}$$

is a continuous bounded homomorphism from $\mathfrak{V}$ *into* $\mathfrak{U}$. *Moreover,* ϑ *is one-to-one if and only if* φ *is onto, and* ϑ *is onto if and only if* φ *is one-to-one.*

Proof. The first task is to show that the function ϑ defined in the statement of the theorem really does map the universe of $\mathfrak{V}$ into the universe of $\mathfrak{U}$. The elements in $\mathfrak{U}$ and in $\mathfrak{V}$ are, by definition, the ultrafilters in $\mathfrak{A}$ and in $\mathfrak{B}$ respectively. The Boolean homomorphism properties of φ imply that if Y is an ultrafilter in $\mathfrak{B}$, then the inverse image of Y under φ, that is to say, the set

$$X = \varphi^{-1}(Y) = \{r \in A : \varphi(r) \in Y\}, \tag{1}$$

is an ultrafilter in $\mathfrak{A}$, and consequently ϑ does map the set V into the set U. In more detail, if r and s belong to X, then the images $\varphi(r)$ and $\varphi(s)$ belong to Y, by (1), and therefore so does the product of these two images. Since

$$\varphi(r \cdot s) = \varphi(r) \cdot \varphi(s),$$

by the homomorphism properties of φ, it follows from (1) that $r \cdot s$ belongs to X. A similar argument shows that if r is in X, and $r \leq s$, then s is in X. If r is not in X, then the image $\varphi(r)$ cannot be in Y, by (1). In this case, $-\varphi(r)$ must be in Y, because Y is an ultrafilter. Since

$$\varphi(-r) = -\varphi(r),$$

by the homomorphism properties of φ, it follows from (1) that $-r$ is in X. Finally, 0 cannot be in X, by (1), because $\varphi(0) = 0$, and 0 is not in Y. Thus, X is an ultrafilter in $\mathfrak{A}$.

The next step is to show that ϑ is a bounded homomorphism. Focus on the case of a ternary relation R that is defined in terms of a binary operator $\circ$. Suppose Y_1, Y_2, and Y_3 are elements in $\mathfrak{V}$ such that

$$R(Y_1, Y_2, Y_3) \tag{2}$$

holds in $\mathfrak{V}$, with the aim of proving that

$$R(\vartheta(Y_1), \vartheta(Y_2), \vartheta(Y_3)) \tag{3}$$

holds in $\mathfrak{U}$. In view of the definition of the relation R in the dual spaces $\mathfrak{V}$ and $\mathfrak{U}$, the hypothesis in (2) is equivalent to the inclusion

$$Y_1 \circ Y_2 \subseteq Y_3, \tag{4}$$

(where the operation on the left side of this inclusion is the complex operation induced on subsets of $\mathfrak{B}$ by the operation $\circ$ in $\mathfrak{B}$), and the proof of (3) amounts to showing that

$$\vartheta(Y_1) \circ \vartheta(Y_2) \subseteq \vartheta(Y_3),$$

or, equivalently, that

$$\varphi^{-1}(Y_1) \circ \varphi^{-1}(Y_2) \subseteq \varphi^{-1}(Y_3) \tag{5}$$

(where the operation on the left sides of these two inclusions is the complex operation induced on subsets of $\mathfrak{A}$ by the operation $\circ$ in $\mathfrak{A}$). To prove (5), consider elements r in $\varphi^{-1}(Y_1)$ and s in $\varphi^{-1}(Y_2)$. The images $\varphi(r)$ and $\varphi(s)$ belong to the sets Y_1 and Y_2 respectively, so the product $\varphi(r) \circ \varphi(s)$ of the two images belongs to the complex product $Y_1 \circ Y_2$, and therefore also to Y_3, by (4). Since

$$\varphi(r \circ s) = \varphi(r) \circ \varphi(s),$$

by the homomorphism properties of φ, it follows that $\varphi(r \circ s)$ is in Y_3, and consequently that $r \circ s$ is in $\varphi^{-1}(Y_3)$, as was to be shown. This completes the proof of (5), and hence also of the implication from (2) to (3).

In order to show that ϑ is also bounded, consider elements X_1 and X_2 in $\mathfrak{U}$, and Y_3 in $\mathfrak{V}$ such that

$$R(X_1, X_2, \vartheta(Y_3)) \tag{6}$$

in $\mathfrak{U}$. Thus,

$$X_1 \circ X_1 \subseteq \vartheta(Y_3) = \varphi^{-1}(Y_3), \tag{7}$$

by the definition of the relation R in $\mathfrak{U}$, and the definition of ϑ. (The operation on the left side in (7) is the complex operation induced on subsets of $\mathfrak{A}$ by the operation $\circ$ in $\mathfrak{A}$.) Elements Y_1 and Y_2 in $\mathfrak{V}$ are to be constructed such that

$$\vartheta(Y_1) = X_1, \qquad \vartheta(Y_2) = X_2, \qquad \text{and} \qquad R(Y_1, Y_2, Y_3). \qquad (8)$$

The construction proceeds stepwise, and the first step is to obtain an ultrafilter Y_1 in $\mathfrak{B}$ with the properties

$$X_1 = \varphi^{-1}(Y_1) \qquad \text{and} \qquad Y_1 \circ \varphi(X_2) \subseteq Y_3. \qquad (9)$$

It is clear from the right-hand inclusion in (9) that we must exclude from Y_1 all elements u in $\mathfrak{B}$ with the property that $u \circ \varphi(r)$ is not in Y_3 for some r in X_2. To this end, write

$$W_1 = \{u \in B : u \circ \varphi(r) \notin Y_3 \text{ for some } r \in X_2\}, \qquad (10)$$

and observe that W_1 is closed under addition. Indeed, let u and v be elements in W_1, and suppose r and s are elements in X_2 such that $u \circ \varphi(r)$ and $v \circ \varphi(s)$ are not in Y_3. Put $t = r \cdot s$, and observe that t also belongs to X_2, since X_2 is an ultrafilter. Moreover,

$$u \circ \varphi(t) \leq u \circ \varphi(r) \qquad \text{and} \qquad v \circ \varphi(t) \leq v \circ \varphi(s),$$

by the homomorphism properties of φ and the monotony of the operator $\circ$. Neither of the terms on the right sides of these inequalities is in Y_3, by assumption, so neither of the terms on the left sides of the inequalities can be in Y_3, by the upward closure of Y_3. It follows that the sum $u \circ \varphi(t) + v \circ \varphi(t)$ also cannot be in Y_3, because Y_3 is an ultrafilter. Since

$$u \circ \varphi(t) + v \circ \varphi(t) = (u + v) \circ \varphi(t),$$

we arrive at the conclusion that $(u + v) \circ \varphi(t)$ cannot be in Y_3, and therefore $u + v$ is in W_1, as claimed. As a consequence of this observation, the set of complements of elements in W_1, that is to say, the set

$$-W_1 - \{-u : u \in W_1\},$$

must be closed under multiplication.

Write $\varphi(X_1)$ for the image of the set X_1 under φ, so that

$$\varphi(X_1) = \{\varphi(r) : r \in X_1\}.$$

We proceed to show that the set

$$\varphi(X_1) \cup -W_1 \tag{11}$$

has the finite meet property. The sets X_1 and $-W_1$ are closed under multiplication, and φ preserves this operation, so it suffices to show that there can be no elements r in X_1 and u in W_1 such that

$$\varphi(r) \cdot -u = 0.$$

Assume, to the contrary, that such elements r and u exist. It follows that $\varphi(r) \leq u$. Since u is in W_1, there must be an element s in X_2 such that $u \circ \varphi(s)$ is not in Y_3, by (10). But then the product $\varphi(r) \circ \varphi(s)$ cannot be in Y_3, because

$$\varphi(r) \circ \varphi(s) \leq u \circ \varphi(s),$$

by the monotony of the operator $\circ$, and the presence of $\varphi(r) \circ \varphi(s)$ in Y_3 would imply that of $u \circ \varphi(s)$, by the upward closure of Y_3. Since

$$\varphi(r \circ s) = \varphi(r) \circ \varphi(s),$$

by the homomorphism properties of φ, it may be concluded that $\varphi(r \circ s)$ is not in Y_3, and therefore that $r \circ s$ is not in $\varphi^{-1}(Y_3)$. However, r is in X_1, and s is in X_2, so $r \circ s$ belongs to the complex product $X_1 \circ X_2$ and therefore also to the inverse image $\varphi^{-1}(Y_3)$, by (7). The desired contradiction has arrived. Conclusion: the set in (11) has the finite meet property.

Every set that has the finite meet property can be extended to an ultrafilter, so there must be an ultrafilter Y_1 in $\mathfrak{B}$ that includes the set in (11). In particular, $\varphi(X_1)$ is included in Y_1, so X_1 is included in the inverse image $\varphi^{-1}(Y_1)$. As both of these last two sets are ultrafilters in $\mathfrak{A}$, it follows that $X_1 = \varphi^{-1}(Y_1)$. The set Y_1 is a proper filter, so no element and its complement can simultaneously belong to Y_1. Since $-W_1$ is included in Y_1, it follows that Y_1 must be disjoint from W_1. This means that if u is in Y_1, then $u \circ \varphi(r)$ belongs to Y_3 for every r in X_2, by (10). Thus, Y_1 possesses the requisite properties stated in (9).

The second step of the construction is to obtain an ultrafilter Y_2 in $\mathfrak{B}$ with the properties

$$X_2 = \varphi^{-1}(Y_2) \quad \text{and} \quad Y_1 \circ Y_2 \subseteq Y_3. \tag{12}$$

The argument is similar to the preceding one: we want $\varphi(X_2)$ to be included in Y_2, and we want to exclude from Y_2 all elements v in $\mathfrak{B}$ with the property that $u \circ v$ is not in Y_3 for some u in Y_1. To this end, write

$$W_2 = \{v \in B : u \circ v \notin Y_3 \text{ for some } u \in Y_1\}. \tag{13}$$

Just as before, one proves that W_2 is closed under addition, and concludes that the set of complements,

$$-W_2 = \{-v : v \in W_2\},$$

is closed under multiplication.

Write $\varphi(X_2)$ for the image of the set X_2 under the mapping φ. We proceed to show that the set

$$\varphi(X_2) \cup -W_2 \tag{14}$$

has the finite meet property. The sets X_2 and $-W_2$ are closed under multiplication, and φ preserves this operation, so it suffices to show, as before, that there are no elements s in X_2 and v in W_2 such that

$$\varphi(s) \cdot -v = 0.$$

Assume, to the contrary, that such elements s and v exist. It follows that $\varphi(s) \le v$. Since v is in W_2, there must be an element u in Y_1 such that $u \circ v$ is not in Y_3, by (13). Consequently, the product $u \circ \varphi(s)$ cannot be in Y_3. In more detail,

$$u \circ \varphi(s) \le u \circ v,$$

by the monotony of the operator $\circ$, so the presence of $u \circ \varphi(s)$ in Y_3 would imply that of $u \circ v$, by the upward closure of Y_3, in contradiction to the assumption that $u \circ v$ is not in Y_3. The failure of $u \circ \varphi(s)$ to be in Y_3 contradicts the right-hand inclusion in (9). Conclusion: the set in (14) has the finite meet property.

It follows from the preceding conclusion that there is an ultrafilter Y_2 in $\mathfrak{B}$ that includes the set in (14). As $\varphi(X_2)$ is included in Y_2, the set X_2 is included in the inverse image $\varphi^{-1}(Y_2)$. These last two sets are ultrafilters, so $X_2 = \varphi^{-1}(Y_2)$. Also, $-W_2$ is included in Y_2, so W_2 is disjoint from Y_2. This means that if v is in Y_2, then $u \circ v$ belongs to Y_3 for every u in Y_1, by (13). Thus, Y_2 possesses the properties required in (12).

The definition of ϑ, and the left-hand equations in (9) and (12), imply the first two equations in (8). The definition of R in $\mathfrak{V}$ and the right-hand inclusion in (12) imply the last part of (8). Thus, ϑ is a bounded homomorphism, as claimed.

Turn now to the task of showing that the mapping ϑ is continuous. The clopen sets in $\mathfrak{U}$ and in $\mathfrak{V}$ are respectively the sets of the form

$$F_r = \{X \in U : r \in X\} \quad \text{and} \quad G_u = \{Y \in V : u \in Y\}$$

for elements r in $\mathfrak{A}$ and u in $\mathfrak{B}$, because $\mathfrak{U}$ and $\mathfrak{V}$ are assumed to be the dual spaces of the algebras $\mathfrak{A}$ and $\mathfrak{B}$ respectively. Every open set in $\mathfrak{U}$ is a union of clopen subsets, and the operation of forming inverse images of sets under a function preserves arbitrary unions. Consequently, it suffices to show that the inverse image under ϑ of every clopen set in $\mathfrak{U}$ is a clopen set in $\mathfrak{V}$. In fact, if F_r is a clopen subset of $\mathfrak{U}$, then

$$\vartheta^{-1}(F_r) = \{Y \in V : \vartheta(Y) \in F_r\} = \{Y \in V : r \in \vartheta(Y)\}$$
$$= \{Y \in V : r \in \varphi^{-1}(Y)\} = \{Y \in V : \varphi(r) \in Y\} = G_{\varphi(r)}, \quad (15)$$

by the definition of the inverse image under ϑ of a set, the definition of F_r, the definition of the function ϑ, the definition of the inverse image under φ of a set, and the definition of $G_{\varphi(r)}$. The set $G_{\varphi(r)}$ is clopen, so ϑ is continuous.

The argument that ϑ is one-to-one if and only if φ is onto is similar to the corresponding argument given in the proof of Theorem 2.16. Statements (1)–(3) from that argument, and the proof of their equivalence, remain unchanged. The argument continues by establishing the equivalence of the following statements: (3) every clopen subset of V is the inverse image under ϑ of some clopen subset of U; (4) for every element u in $\mathfrak{B}$ there is an element r in $\mathfrak{A}$ such that $G_u = \vartheta^{-1}(F_r)$; (5) for every element u in $\mathfrak{B}$ there is an element r in $\mathfrak{A}$ such that $G_u = G_{\varphi(r)}$; (6) φ is onto. The equivalence of (3) and (4) is obvious from the description of the clopen subsets of U and V given above; the equivalence of (4) and (5) follows from (15); and the equivalence of (5) and (6) is a consequence of the fact that the correspondence mapping each element u in $\mathfrak{B}$ to the clopen set G_u is a bijection from $\mathfrak{B}$ to the set of clopen subsets of $\mathfrak{V}$. Conclusion: ϑ is one-to-one if and only if φ is onto.

The argument that ϑ is onto if and only if φ is one-to-one is also similar to the corresponding argument given in the proof of Theorem 2.16.

Statements (1)–(4) from that argument, and the proof of their equivalence, remain unchanged. The argument continues by establishing the equivalence of the following statements, where we use (15) to write $G_{\varphi(r)}$ for the inverse image $\vartheta^{-1}(F_r)$: (4) if F is a clopen subset of U such that $\vartheta^{-1}(F)$ is empty, then F is empty; (5) if $G_{\varphi(r)}$ is empty, then F_r is empty; (6) if $\varphi(r) = 0$, then $r = 0$; (7) φ is one-to-one. Statement (5) is just a rephrasing of assertion (4) using the introduced notation. The equivalence of (5) and (6) follows from the fact that the only clopen subsets of U and V that are empty are respectively the sets F_r and G_u with $r = 0$ and $u = 0$, by the monomorphism properties of the canonical embeddings. The equivalence of (6) and (7) is just the assertion that a Boolean homomorphism is one-to-one if and only if its kernel contains only the zero element. Conclusion: ϑ is onto if and only if φ is one-to-one. $\square$

The assertion in Theorem 2.18 that, under the hypotheses of the theorem, the mapping ϑ is a continuous bounded homomorphism, and the proof of this assertion, are due to Goldblatt [13] (see Theorem 2.3.2 and its proof in [13]). He also points out that if φ is onto or one-to-one, then ϑ is one-to-one or onto respectively. (An earlier version of Goldblatt's result, for Boolean algebras with a single unary operator, is given in Theorem 10.9 of Goldblatt [12].)

Theorem 2.18 also has a version that refers not to Boolean algebras with operators and their dual spaces, but rather to relational spaces and their dual algebras.

Corollary 2.19. *Let $\mathfrak{U}$ and $\mathfrak{V}$ be relational spaces, and $\mathfrak{A}$ and $\mathfrak{B}$ their respective dual algebras. If φ is a homomorphism from $\mathfrak{A}$ to $\mathfrak{B}$, then the function δ defined on elements s in $\mathfrak{V}$ by*

$$\delta(s) = r \qquad \textit{if and only if} \qquad \varphi^{-1}(Y_s) = X_r,$$

where

$$X_r = \{F \in A : r \in F\} \qquad \textit{and} \qquad Y_s = \{G \in B : s \in G\},$$

is a continuous bounded homomorphism from $\mathfrak{V}$ into $\mathfrak{U}$. Moreover, δ is one-to-one if and only if φ is onto, and δ is onto if and only if φ is one-to-one.

Proof. Let $\mathfrak{U}^*$ and $\mathfrak{V}^*$ be the dual spaces of the dual algebras $\mathfrak{A}$ and $\mathfrak{B}$ respectively. Thus, $\mathfrak{U}^*$ is the second dual of $\mathfrak{U}$, and $\mathfrak{V}^*$ the second dual of $\mathfrak{V}$, by the definition of the second duals. The functions ϑ_1 and ϑ_2 defined by

$$\vartheta_1(r) = X_r \quad \text{and} \quad \vartheta_2(s) = Y_s \tag{1}$$

for r in $\mathfrak{U}$ and s in $\mathfrak{V}$ are homeo-isomorphisms from $\mathfrak{U}$ to $\mathfrak{U}^*$ and from $\mathfrak{V}$ to $\mathfrak{V}^*$ respectively, by Theorem 2.11.

Assume that φ is a homomorphism from $\mathfrak{A}$ to $\mathfrak{B}$. The function ϑ defined on elements Y in $\mathfrak{V}^*$ by

$$\vartheta(Y) = \varphi^{-1}(Y) = \{F \in A : \varphi(F) \in Y\} \tag{2}$$

is a continuous bounded homomorphism from $\mathfrak{V}^*$ to $\mathfrak{U}^*$, by the first part of Theorem 2.18 (with $\mathfrak{U}^*$ and $\mathfrak{V}^*$ in place of $\mathfrak{U}$ and $\mathfrak{V}$, and F in place of r). Moreover, ϑ is one-to-one or onto if and only if φ is onto or one-to-one respectively, by the second part of Theorem 2.18. The composition

$$\delta = \vartheta_1^{-1} \circ \vartheta \circ \vartheta_2 \tag{3}$$

is therefore a continuous bounded homomorphism from $\mathfrak{V}$ to $\mathfrak{U}$ (see the diagram below). Moreover, since ϑ_1 and ϑ_2 are bijections, the mapping δ is one-to-one or onto if and only the mapping ϑ is one-to-one or onto respectively. Consequently, δ is one-to-one or onto if and only φ is onto or one-to-one respectively.

$$
\begin{array}{ccc}
\mathfrak{V} & \xrightarrow{\ \delta\ } & \mathfrak{U} \\
\vartheta_2 \downarrow & & \downarrow \vartheta_1 \\
\mathfrak{V}^* & \xrightarrow{\ \vartheta\ } & \mathfrak{U}^*
\end{array}
$$

If $\varphi^{-1}(Y_s) = X_r$, then $\vartheta(Y_s) = X_r$, by (2), and therefore

$$\delta(s) = (\vartheta_1^{-1} \circ \vartheta \circ \vartheta_2)(s) = \vartheta_1^{-1}(\vartheta(\vartheta_2(s)))$$
$$= \vartheta_1^{-1}(\vartheta(Y_s)) = \vartheta_1^{-1}(X_r) = r,$$

by (3) and (1). On the other hand, if $\delta(s) = r$, then

$$\vartheta(Y_s) = (\vartheta_1 \circ \delta \circ \vartheta_2^{-1})(Y_s) = \vartheta_1(\delta(\vartheta_2^{-1}(Y_s)))$$
$$= \vartheta_1(\delta(s)) = \vartheta_1(r) = X_r,$$

by (3) and (1), and therefore $\varphi^{-1}(Y_s) = X_r$, by (2). Thus, the mapping δ is determined by the equivalence stated in the corollary. $\quad\square$

The function δ in Corollary 2.19 is called the (*first*) *dual*, or the *dual continuous bounded homomorphism*, of the homomorphism φ. The equivalence defining this dual in the statement of the corollary can be formulated in a different way. The right side of this equivalence, namely the equation

$$\varphi^{-1}(Y_s) = X_r,$$

expresses that

$$F \in X_r \qquad \text{if and only if} \qquad \varphi(F) \in Y_s$$

for every element F in $\mathfrak{A}$ (that is to say, for every clopen subset F of $\mathfrak{U}$), or equivalently that

$$r \in F \qquad \text{if and only if} \qquad s \in \varphi(F)$$

for every element F in $\mathfrak{A}$, by the definitions of the ultrafilters X_r and Y_s, and the definition of the inverse image $\varphi^{-1}(Y_s)$. Consequently the function δ defined in the corollary is completely determined by the equivalence

$$\delta(s) \in F \qquad \text{if and only if} \qquad s \in \varphi(F)$$

for all elements s in $\mathfrak{V}$ and F in $\mathfrak{A}$. This equivalence may, in turn, be reformulated as

$$s \in \delta^{-1}(F) \qquad \text{if and only if} \qquad s \in \varphi(F)$$

for all elements s in $\mathfrak{V}$ and F in $\mathfrak{A}$. This last equivalence simply says that the dual δ is determined by the validity of the equation

$$\varphi(F) = \delta^{-1}(F)$$

for every F in $\mathfrak{A}$. This equation is of course the definition of φ in terms of δ, by Theorem 2.16. The point is that the equation is also valid in the context of Corollary 2.19, where we are defining δ in terms of φ. Notice in passing that in view of the second equivalence above, the definition of δ in Corollary 2.19 may be written in the form

$$\delta(s) = r \qquad \text{if and only if} \qquad r \in \bigcap\{F : s \in \varphi(F)\}.$$

We continue with the assumptions of the corollary, namely that $\mathfrak{U}$ and $\mathfrak{V}$ are relational spaces with dual algebras $\mathfrak{A}$ and $\mathfrak{B}$ respectively.

If φ is a homomorphism from $\mathfrak{A}$ to $\mathfrak{B}$, and δ the dual continuous bounded homomorphism from $\mathfrak{V}$ to $\mathfrak{U}$, then the dual of δ is called the *second dual* of φ. This second dual is a homomorphism φ^* from $\mathfrak{A}$ to $\mathfrak{B}$ that is defined on elements F in $\mathfrak{A}$ by

$$\varphi^*(F) = \delta^{-1}(F),$$

by Theorem 2.16 (with δ and φ^* in place of ϑ and φ respectively). Comparing this definition with the conclusion of the last paragraph, we see that

$$\varphi^*(F) = \delta^{-1}(F) = \varphi(F).$$

In other words, φ is its own second dual. Similarly, if ϑ is a continuous bounded homomorphism from $\mathfrak{V}$ to $\mathfrak{U}$, and φ the dual homomorphism from $\mathfrak{A}$ to $\mathfrak{B}$, then the dual of φ is called the *second dual* of ϑ. This second dual is a function ϑ^* from $\mathfrak{V}$ to $\mathfrak{U}$ that is determined by the equivalence

$$\vartheta^*(s) \in F \qquad \text{if and only if} \qquad s \in \varphi(F)$$

for all elements s in $\mathfrak{V}$ and F in $\mathfrak{A}$, by the remarks of the preceding paragraph (with ϑ^* in place of δ). On the other hand, since φ is the dual of ϑ, we have

$$s \in \varphi(F) \qquad \text{if and only if} \qquad \vartheta(s) \in F$$

for all s in $\mathfrak{V}$ and F in $\mathfrak{A}$, by the remarks following Corollary 2.17 (with s in place of u). Thus,

$$\vartheta^*(s) \in F \qquad \text{if and only if} \qquad \vartheta(s) \in F$$

for all s in $\mathfrak{V}$ and F in $\mathfrak{A}$, that is to say, for all points s in $\mathfrak{V}$ and all clopen subsets F of $\mathfrak{U}$. Fix the point s for a moment. The space $\mathfrak{U}$ is Boolean, so distinct points are separated by clopen sets. It follows that the image points $\vartheta^*(s)$ and $\vartheta(s)$ must be the same, for otherwise they would be separated by a clopen subset F of $\mathfrak{U}$. Since this is true for every element s in $\mathfrak{V}$, the function ϑ coincides with its own second dual.

Suppose next that $\mathfrak{U}$, $\mathfrak{V}$, and $\mathfrak{W}$ are relational spaces with dual algebras $\mathfrak{A}$, $\mathfrak{B}$, and $\mathfrak{C}$ respectively. Let δ be a continuous bounded homomorphism from $\mathfrak{W}$ to $\mathfrak{V}$, and ϑ a continuous bounded homomorphism from $\mathfrak{V}$ to $\mathfrak{U}$. It is easy to check that the composition $\vartheta \circ \delta$ is a continuous bounded homomorphism from $\mathfrak{W}$ to $\mathfrak{U}$. The dual of δ is the homomorphism ϱ from $\mathfrak{B}$ to $\mathfrak{C}$ defined by

$$\varrho(G) = \delta^{-1}(G)$$

for elements G in $\mathfrak{B}$, and the dual of ϑ is the homomorphism φ from $\mathfrak{A}$ to $\mathfrak{B}$ defined by

$$\varphi(F) = \vartheta^{-1}(F)$$

for elements F in $\mathfrak{A}$, by Theorem 2.16 and the definition of the dual of a continuous bounded homomorphism. The composition $\varrho \circ \varphi$ is a homomorphism from $\mathfrak{A}$ to $\mathfrak{C}$, and

$$(\varrho \circ \varphi)(F) = \varrho(\varphi(F)) = \delta^{-1}(\vartheta^{-1}(F)) = (\vartheta \circ \delta)^{-1}(F)$$

for all F in $\mathfrak{A}$. Consequently, the composition $\varrho \circ \varphi$ is the dual of the composition $\vartheta \circ \delta$, by the definition of that dual.

The next theorem summarizes the results of this section.

Theorem 2.20. *Let* $\mathfrak{U}$, $\mathfrak{V}$, *and* $\mathfrak{W}$ *be relational spaces, and* $\mathfrak{A}$, $\mathfrak{B}$, *and* $\mathfrak{C}$ *their respective dual algebras. There is a bijective correspondence between the set of continuous bound homomorphisms* ϑ *from* $\mathfrak{V}$ *to* $\mathfrak{U}$ *and the set of homomorphisms* φ *from* $\mathfrak{A}$ *to* $\mathfrak{B}$ *such that the equivalence*

$$u \in \varphi(F) \qquad \textit{if and only if} \qquad \vartheta(u) \in F$$

holds for all sets F *in* $\mathfrak{A}$ *and all elements* u *in* $\mathfrak{V}$. *Each of the mappings* ϑ *and* φ *is its own second dual. The continuous bounded homomorphism* ϑ *is one-to-one if and only if its dual homomorphism* φ *is onto, and* ϑ *is onto if and only if* φ *is one-to-one. If* δ *is a continuous bounded homomorphism from* $\mathfrak{W}$ *to* $\mathfrak{V}$, *with dual* ϱ, *then the dual of the composition* $\vartheta \circ \delta$ *is the composition* $\varrho \circ \varphi$.

Part of the contents of the preceding theorem may be expressed by saying that the correspondence taking each relational space to its dual algebra, and each continuous bounded homomorphism to its dual homomorphism, is a contravariant functor from the category of all relational spaces with continuous bounded homomorphisms as morphisms to the category of all Boolean algebras with operators with homomorphisms as morphisms; and the correspondence taking each Boolean algebra with operators to its dual space, and each homomorphism to its dual continuous bounded homomorphism, is a contravariant functor from the category of all Boolean algebra with operators with homomorphisms as morphisms to the category of all relational spaces with continuous bounded homomorphisms as morphisms; and these two contravariant functors are inverses of one another. Consequently, the two categories are dually equivalent. This last statement is contained in Theorem 2.3.3 of Goldblatt [13]; an earlier version of his theorem, applicable to the categories of modal algebras and descriptive modal frames, is given at the end of Section 10 in Goldblatt [12].

2.5 Duality for Ideals

The duality between homomorphisms and continuous bounded homomorphisms implies a duality between ideals (see Definition 1.15) and open sets: every ideal has a dual open set with a special property, and every open set with this special property has a dual ideal. The special property that the open sets possess is set forth in the next definition.

Definition 2.21. An open subset H of a relational space $\mathfrak{U}$ is called *special* if for every relation R in $\mathfrak{U}$ of rank $n + 1$ (with $n \geq 1$), and every sequence of clopen subsets $F_0, \ldots, F_{n-1}$ of $\mathfrak{U}$, if F_i is included in H for some $i < n$, then the clopen image set

$$R^*(F_0 \times \cdots \times F_{n-1}) = \{t \in U : R(r_0, \ldots, r_{n-1}, t)$$
$$\text{for some } r_0 \in F_0, \ldots, r_{n-1} \in F_{n-1}\}$$

is included in H. $\square$

For example, if $\mathfrak{U}$ has a single ternary relation R, then an open subset H of $\mathfrak{U}$ is special provided that for every pair of clopen subsets F and G of $\mathfrak{U}$, if one of F and G is included in H, then the clopen image set

$$R^*(F \times G) = \{t \in U : R(r, s, t) \text{ for some } r \in F \text{ and } s \in G\}$$

is included in H.

Suppose $\mathfrak{U}$ is a relational space, and $\mathfrak{A}$ its dual algebra. If M is an ideal in $\mathfrak{A}$, then the union of the clopen sets in M is a special open set in $\mathfrak{U}$. Conversely, if H is an arbitrary special open set in $\mathfrak{U}$, then the set of all clopen subsets of H is an ideal in $\mathfrak{A}$. In order to prove these assertion, it is helpful to make a preliminary observation: if M is a set of elements in $\mathfrak{A}$—that is to say, if M is a set of clopen subsets of $\mathfrak{U}$— and if M satisfies conditions (ii) and (iii) in the definition of an ideal, then an element F in $\mathfrak{A}$ belongs to M just in case F is included in the union of M. One direction of this observation is clear: if F belongs to M, then F is certainly included in the union of all of the sets in M. To prove the reverse direction, suppose F is included in the union of M. The set F, being clopen, is closed and therefore compact in the topology of $\mathfrak{U}$, and the sets in M, being clopen, are open and therefore form an open cover of F (because F is included in their union). Apply compactness to obtain a finite subset M_0 of M such that the union of the sets in M_0 includes F. The set M is closed under finite unions, by condition (ii), so the union of the sets in M_0 is a clopen set G that belongs to M. The set M is also downward closed, by condition (iii), and F is included in G, so F must belong to M.

Lemma 2.22. *Let $\mathfrak{U}$ be a relational space and $\mathfrak{A}$ its dual algebra. If M is an ideal in $\mathfrak{A}$, then the union of the sets in M is a special open subset of $\mathfrak{U}$. Inversely, if H is a special open set in $\mathfrak{U}$, then the set of clopen sets in $\mathfrak{U}$ that are included in H is an ideal in $\mathfrak{A}$.*

Proof. Suppose first that M is an ideal in $\mathfrak{A}$, and let H be the union of the sets in M. Since $\mathfrak{A}$ is the dual algebra of $\mathfrak{U}$, the elements in $\mathfrak{A}$ are the clopen subsets of $\mathfrak{U}$. In particular, M is a set of clopen subsets of $\mathfrak{U}$, so its union H is an open set in $\mathfrak{U}$. To show that H is special, consider the case of a ternary relation R in $\mathfrak{U}$ and the binary operator $\circ$ that is defined in $\mathfrak{A}$ in terms of R. Let F and G be clopen sets in $\mathfrak{U}$, and suppose that F is included in H. The set F must then belong to the ideal M, by the observation preceding the lemma, and therefore the set $F \circ G$ also belongs to M, by condition (iv) in the definition of an ideal. The set $F \circ G$ coincides with the image set $R^*(F \times G)$, by the definition of the dual algebra $\mathfrak{A}$ (see Lemma 2.9 and its proof), so the image set belongs to M and is therefore included in the union H. A similar argument applies if G is included in H. Consequently, H is a special open set, by Definition 2.21.

To prove the second assertion of the lemma, assume that H is an arbitrary special open set in $\mathfrak{U}$, and let M be the set of all clopen subsets of H. It must be shown that M satisfies conditions (i)–(iv) in the definition of an ideal. The empty set is obviously a clopen subset of $\mathfrak{U}$ that is included in H, so the empty set belongs to M. Thus, condition (i) is satisfied. The union of two clopen subsets of $\mathfrak{U}$ that are included in H is again a clopen subset of $\mathfrak{U}$ that is included in H. Consequently, M contains the union of any two of its elements and therefore satisfies condition (ii). The intersection of a clopen subset of $\mathfrak{U}$ that is included in H with an arbitrary clopen subset of $\mathfrak{U}$ is again a clopen subset of $\mathfrak{U}$ that is included in H, so M satisfies condition (iii). To show that M satisfies condition (iv), consider the case of a binary operator $\circ$ in $\mathfrak{A}$ that is defined in terms of a ternary relation R in $\mathfrak{U}$. Let F and G be clopen sets in $\mathfrak{U}$, and suppose F belongs to M. The clopen image set $R^*(F \times G)$ is then included in H, by Definition 2.21 and the assumption that H is a special open set, so this image set must belong to M, by the observation made before the lemma. Since this image set coincides with $F \circ G$, it follows that the latter must belong to M. A similar argument applies if G belongs to M. Consequently, M satisfies conditions (iv) and is therefore an ideal in $\mathfrak{A}$. □

If M is an ideal in the dual algebra $\mathfrak{A}$ of a relational space $\mathfrak{U}$, then the special open set that is the union of the sets in M is called the (*first*) *dual*, or the *dual open set*, of M. Similarly, if H is a special open set in $\mathfrak{U}$, then the ideal of clopen subsets of $\mathfrak{U}$ that are included in H is called the (*first*) *dual*, or the *dual ideal*, of H.

If M is the dual ideal of an arbitrary special open set H in $\mathfrak{U}$, then M is, by definition, the set of clopen subsets of H. The dual open set of M is, by definition, the union of M. This union must coincide with H, because every open set in a Boolean space is the union of its clopen subsets. It follows that the second dual of every special open set in $\mathfrak{U}$ is itself. Similarly, if H is the dual open set of an arbitrary ideal M in $\mathfrak{A}$, then H is, by definition, the union of M. The dual ideal of H is, by definition, the set of all clopen subsets of H. This ideal must coincide with M, because a clopen set is included in H if and only if it belongs to M, by the observation made before the preceding lemma. It follows that the second dual of every ideal in $\mathfrak{A}$ is itself.

If M and N are ideals in $\mathfrak{A}$, and if H and K are their respective dual special open sets, then

$$M \subseteq N \qquad \text{if and only if} \qquad H \subseteq K.$$

The implication from left to right is clear, since H and K are defined to be the unions of M and N respectively. On the other hand, if H is included in K, then every clopen subset of H is also a clopen subset of K. Consequently, M is included in N, because M and N are defined to be the sets of all clopen subsets of H and K respectively.

The principal facts about the duality between ideals and special open sets are summarized in the following duality theorem for ideals.

Theorem 2.23. *The dual of every special open subset of a relational space $\mathfrak{U}$ is an ideal in the dual algebra $\mathfrak{A}$, and the dual of every ideal in $\mathfrak{A}$ is a special open subset of $\mathfrak{U}$. The second dual of every ideal and of every special open set is itself. The function that maps every ideal in $\mathfrak{A}$ to its dual special open set is an isomorphism from the lattice of ideals in $\mathfrak{A}$ to the lattice of special open sets in $\mathfrak{U}$.*

It is illuminating to look at the duality between ideals and special open sets from the perspective of an arbitrary Boolean algebra with operators $\mathfrak{A}$ and its dual space $\mathfrak{U}$, instead of from the perspective of an arbitrary relational space $\mathfrak{U}$ and its dual algebra $\mathfrak{A}$, as in Theorem 2.23. The elements in $\mathfrak{U}$ are the ultrafilters in $\mathfrak{A}$, and every element r in $\mathfrak{A}$ is identified via the canonical isomorphism with an element in the second dual of $\mathfrak{A}$—that is to say, in the dual algebra of $\mathfrak{U}$—namely with the clopen set

$$F_r = \{X \in U : r \in X\}.$$

Every ideal M is $\mathfrak{A}$ is consequently identified with the ideal of clopen sets

$$M_0 = \{F_r : r \in M\}$$

in the second dual. The ideal M determines a special open set in $\mathfrak{U}$, namely the union

$$F_M = \bigcup \{F_r : r \in M\}$$

of the clopen sets belonging to M_0. Furthermore, every special open set H in $\mathfrak{U}$ has the form $H = F_M$ for some ideal M in $\mathfrak{A}$. Indeed, if M_0 is taken to be the set of clopen subsets of H, then M_0 is an ideal in the dual of $\mathfrak{U}$, and the union of the sets in M_0 is just the special open set H, by the preceding theorem. If M is the ideal in $\mathfrak{A}$ that corresponds to M_0 (under the canonical isomorphism), then

$$H = \bigcup M_0 = \bigcup \{F_r : r \in M\} = F_M.$$

Thus, the special open sets in $\mathfrak{U}$ are precisely the sets F_M, where M ranges over the ideals in $\mathfrak{A}$.

The canonical isomorphism from $\mathfrak{A}$ to its second dual obviously induces an isomorphism between the corresponding lattices of ideals. The lattice of ideals of the second dual is isomorphic to the lattice of special open subsets of $\mathfrak{U}$, by Theorem 2.23. Consequently, the correspondence that maps each ideal M in $\mathfrak{A}$ to the special open set F_M in $\mathfrak{U}$ is an isomorphism from the lattice of ideals in $\mathfrak{A}$ to the lattice of special open sets in $\mathfrak{U}$. In this formulation of the duality between ideals and special open sets, the assertion that the second dual of every ideal is itself is not literally true; one must first identify the algebra $\mathfrak{A}$ with its second dual before the assertion becomes true.

Theorem 2.23 has an analogue for filters. (The notion of a filter in a Boolean algebra with operators is defined in a dual manner to that of an ideal.) Let $\mathfrak{U}$ be a relational space, and $\mathfrak{A}$ the dual algebra of $\mathfrak{U}$. A closed subset of $\mathfrak{U}$ is called *special* if it is the complement of a special open set. If N is a filter in $\mathfrak{A}$, then the intersection of the clopen sets in N is a special closed subset of $\mathfrak{U}$; in fact, if M is the ideal that is the dual of N, then the intersection of N is the complement of the special open set that is the union of M. Call this special closed set the (*first*) *dual*, or the *dual closed subset*, of N. Conversely, if H is a special closed subset of $\mathfrak{U}$, then the set of all clopen sets in $\mathfrak{U}$ that include H is a filter in $\mathfrak{A}$. Call this filter the (*first*) *dual*, or the *dual filter*, of H. Each filter in $\mathfrak{A}$ and each special closed set in $\mathfrak{U}$ is its own second dual, and the function that maps each filter in $\mathfrak{A}$ to its dual special closed set is an isomorphism from the lattice of filters in $\mathfrak{A}$ to the lattice of special closed sets in $\mathfrak{U}$. A related result, discovered independently by Celani, is given in Proposition 29 of [2].

2.6 Duality for Quotients

The duality between the special open subsets of a relational space $\mathfrak{U}$ and the ideals in the dual algebra $\mathfrak{A}$ implies a duality between quotients of $\mathfrak{A}$ and certain subspaces of $\mathfrak{U}$. We begin by clarifying the relationship between inner subuniverses (see Definition 1.12 and the remarks following it) and special closed subsets.

Lemma 2.24. *A closed subset V of a relational space $\mathfrak{U}$ is a inner subuniverse of $\mathfrak{U}$ if and only if V is a special closed subset of $\mathfrak{U}$.*

Proof. Focus on the case of a fundamental ternary relation R. Assume first that V is a special closed subset of $\mathfrak{U}$, and observe that the complement $\sim V$ is, by definition, a special open set. To check that V is an inner subuniverse of $\mathfrak{U}$, consider elements r, s, and t in $\mathfrak{U}$ such that $R(r, s, t)$ holds in $\mathfrak{U}$. If at least one of r and s is not in V, then t cannot be in V. Indeed, suppose r is not in V. The clopen sets form a base for the topology of $\mathfrak{U}$, and $\sim V$ is an open set that contains r, so there must be a clopen set F in $\mathfrak{U}$ such that r is in F and F is included in $\sim V$. The set U is also clopen, so the image clopen set

$$R^*(F \times U) = \{w : R(p, q, w) \text{ for some } p \in F \text{ and } q \in U\}$$

is included in $\sim V$, by the assumption that $\sim V$ is a special open set. In particular, since r is in F, and s in U, and since $R(r, s, t)$ holds, it may be concluded that t is in $R^*(F \times U)$ and therefore also in $\sim V$, so t cannot be in V. An analogous argument applies if s is not in V. Conclusion: if $R(r, s, t)$ holds in $\mathfrak{U}$, and if t is in V, then r and s must both belong to V. Consequently, V is an inner subuniverse of $\mathfrak{U}$.

Suppose now that V is an inner subuniverse of $\mathfrak{U}$ that is closed in the topology of $\mathfrak{U}$. The complement $\sim V$ is then an open subset of $\mathfrak{U}$, and it must be shown that this complement is special. To this end, consider clopen sets F and G, and suppose that F is included in $\sim V$. If t belongs to the image set $R^*(F \times G)$, then there must be elements r in F and s in G such that $R(r, s, t)$ is true in $\mathfrak{V}$, by the definition of the image set. Since V is an inner subuniverse of $\mathfrak{U}$, the presence of t in V would force both r and s to belong to V. But r belongs to F, which is included in $\sim V$. Consequently, t cannot be in V, so t belongs to $\sim V$. This is true for every element t in the set $R^*(F \times G)$, so the entire set is included in $\sim V$. A similar argument applies if G is included in $\sim V$, so $\sim V$ is a special open set. $\square$

By a *restriction* of a relational space $\mathfrak{U}$ to a subset V, we understand the relational structure whose universe is V and whose fundamental relations are the restrictions to V of the fundamental relations in $\mathfrak{U}$, together with the topology that V inherits from $\mathfrak{U}$. In general, the restriction of a relational space to a subset is not a relational space. The next lemma gives a sufficient condition for such a restriction to be a relational space.

Lemma 2.25. *If V is a closed subset and an inner subuniverse of a relational space $\mathfrak{U}$, then the restriction of $\mathfrak{U}$ to V is a relational space.*

Proof. The inherited topology turns V into a Boolean space, because V is assumed to be a closed subset of the Boolean space U (see Lemma 2 on p. 306 of [10]). Let $\mathfrak{V}$ be the restriction of $\mathfrak{U}$ to the set V. It must be shown that the relations in $\mathfrak{V}$ are clopen, and those of rank at least two are continuous, in the sense of Definition 2.2 and the remarks preceding it. Focus on the case of a ternary relation R.

Suppose F and G are clopen subsets of $\mathfrak{V}$, with the aim of proving that the image set

$$R^*(F \times G) = \{t \in V : R(r, s, t) \text{ for some } r \in F \text{ and } s \in G\} \quad (1)$$

is clopen. (Since the fundamental relations of $\mathfrak{V}$ are the restrictions to V of the fundamental relations of $\mathfrak{U}$, it does not matter whether the relation R in (1) is viewed as the ternary relation in $\mathfrak{V}$ or as the ternary relation in $\mathfrak{U}$.) The set V is assumed to be closed, so there must be clopen subsets $\bar{F}$ and $\bar{G}$ of $\mathfrak{U}$ such that

$$F = \bar{F} \cap V \quad \text{and} \quad G = \bar{G} \cap V \quad (2)$$

(see Lemma 2 on p. 306 of [10]). Because $\bar{F}$ and $\bar{G}$ are clopen in $\mathfrak{U}$, the image set

$$R^*(\bar{F} \times \bar{G}) = \{t \in U : R(r, s, t) \text{ for some } r \in \bar{F} \text{ and } s \in \bar{G}\}. \quad (3)$$

is clopen in $\mathfrak{U}$, by Definition 2.2 and the assumption that $\mathfrak{U}$ is a relational space.

Observe that

$$R^*(F \times G) = R^*(\bar{F} \times \bar{G}) \cap V. \quad (4)$$

Indeed, if t belongs to the left side of (4), then $R(r, s, t)$ holds in $\mathfrak{V}$ for some r in F and s in G, by (1). In particular, t is in V, by (2). Since F is included in $\bar{F}$, and G in $\bar{G}$, by (2), the element r belongs to $\bar{F}$, and s to $\bar{G}$, and therefore t belongs to the right side of (4), by (3) and the assumption that V is a subset of $\mathfrak{U}$. On the other hand, if t belongs to the right side of (4), then t is in V, and $R(r, s, t)$ holds in $\mathfrak{U}$ for some r in $\bar{F}$ and s in $\bar{G}$. Since V is an inner subuniverse of $\mathfrak{U}$, and t is in V, the elements r and s must be in V (see the remarks following Definition 1.12). Consequently, r is in F and s in G, by (2), so t belongs to the left side of (4), by (1).

The equation in (4) shows that $R^*(F \times G)$ is the intersection with V of a clopen subset of $\mathfrak{U}$. It follows that $R^*(F \times G)$ is a clopen subset of $\mathfrak{V}$. Since this is true for every pair of clopen sets F and G in $\mathfrak{V}$, the relation R in $\mathfrak{V}$ is clopen.

The next task is to prove that the relation R in $\mathfrak{V}$ is weakly continuous in the sense of Definition 2.12. To this end, fix an arbitrary element t in $\mathfrak{V}$, with the aim of showing that the set

$$H_t = \{(r,s) : r, s \in V \text{ and } R(r,s,t)\} \tag{5}$$

is closed in the product space $V \times V$. The set

$$\bar{H}_t = \{(r,s) : r, s \in U \text{ and } R(r,s,t)\} \tag{6}$$

is closed in the product space $U \times U$, by Theorem 2.13, because $\mathfrak{U}$ is assumed to be a relational space. It is not difficult to check that

$$H_t = \bar{H}_t \cap (V \times V). \tag{7}$$

Indeed, if a pair (r,s) belongs to the left side of (7), then $R(r,s,t)$ holds in $\mathfrak{V}$, by (5). It follows that $R(r,s,t)$ holds in $\mathfrak{U}$, because the fundamental relations in $\mathfrak{V}$ are the restrictions to V of the fundamental relations in $\mathfrak{U}$; and consequently, the pair (r,s) belongs to the right side of (7), by (6). On the other hand, if a pair (r,s) belongs to the right side of (7), then the elements r and s are in V, as is t (by assumption), and $R(r,s,t)$ holds in $\mathfrak{U}$, by (6). It follows that $R(r,s,t)$ holds in $\mathfrak{V}$, because the fundamental relations in $\mathfrak{V}$ are the restrictions to V of the fundamental relations in $\mathfrak{U}$; and the pair (r,s) therefore belongs to the left side of (7), by (5).

The equation in (7) shows that H_t is the intersection with $V \times V$ of a closed subset of $U \times U$. Since V is assumed to be closed in $\mathfrak{U}$, it follows that H_t is a closed subset of $V \times V$. This is true for every element t in $\mathfrak{V}$, so the relation R in $\mathfrak{V}$ is weakly continuous. It has been shown that the restriction $\mathfrak{V}$ is a Boolean space and that the fundamental relations in $\mathfrak{V}$ are clopen, and those of rank at least two are weakly continuous, under the inherited topology. Apply Theorem 2.14 to conclude that $\mathfrak{V}$ is a relational space. $\square$

In order to state the version of the sub-quotient duality theorem that applies to quotients of Boolean algebras with operators, it is necessary to define the appropriate notion of a subspace for relational spaces.

Definition 2.26. A relational space $\mathfrak{V}$ is an *inner subspace* of a relational space $\mathfrak{U}$ if algebraically $\mathfrak{V}$ is an inner substructure of $\mathfrak{U}$ and if the topology on $\mathfrak{V}$ is the subspace topology inherited from $\mathfrak{U}$. $\square$

Notice that every inner subspace $\mathfrak{V}$ of a relational space $\mathfrak{U}$ must be the restriction of $\mathfrak{U}$ to some subset of $\mathfrak{U}$, namely the subset V that is the universe of $\mathfrak{V}$, by the definition of an inner substructure, the definition of the subspace topology, and the definition of a restriction. The next theorem characterizes those subsets of $\mathfrak{U}$ that lead to inner subspaces.

Theorem 2.27. *For a subset V of a relational space $\mathfrak{U}$, the following conditions are equivalent.*

(i) *V is a special closed subset of $\mathfrak{U}$.*
(ii) *V is a closed subset and an inner subuniverse of $\mathfrak{U}$.*
(iii) *The restriction of $\mathfrak{U}$ to V is an inner subspace of $\mathfrak{U}$.*

Proof. The equivalence of (i) and (ii) follows at once from Lemma 2.24. For the implication from (ii) to (iii), assume that V is a closed subset and an inner subuniverse of $\mathfrak{U}$. The restriction of $\mathfrak{U}$ to V is then a relational space $\mathfrak{V}$, by Lemma 2.25. Algebraically, $\mathfrak{V}$ is an inner substructure of $\mathfrak{U}$, by the assumption that V is an inner subuniverse of $\mathfrak{U}$. Topologically, $\mathfrak{V}$ is a subspace of $\mathfrak{U}$, by the definition of a restriction of a relational space. Consequently, $\mathfrak{V}$ is an inner subspace of $\mathfrak{U}$, by Definition 2.26.

To establish the reverse implication from (iii) to (ii), assume that $\mathfrak{V}$ is an inner subspace of $\mathfrak{U}$. The universe of $\mathfrak{V}$ must then be an inner subuniverse of $\mathfrak{U}$, by Definition 2.26. Topologically, $\mathfrak{V}$ is a subspace of $\mathfrak{U}$, by Definition 2.26, and the universe of $\mathfrak{V}$ is compact because $\mathfrak{V}$ is assumed to be a relational space. Consequently, this universe must be closed in the topology of $\mathfrak{U}$, because a subset of a compact Hausdorff space is compact if and only if it is closed (see pp. 271–272 of [10]).   □

Corollary 2.28. *An inner subuniverse V of a relational space $\mathfrak{U}$ is the universe of an inner subspace of $\mathfrak{U}$ if and only if V is closed in the topology of $\mathfrak{U}$.*

The duality between quotient algebras and inner subspaces may now be formulated as follows.

Theorem 2.29. *There is a bijective correspondence between the inner subspaces of a relational space $\mathfrak{U}$ and the quotients of its dual algebra $\mathfrak{A}$. If $\mathfrak{V}$ is an inner subspace of $\mathfrak{U}$, then the dual algebra of $\mathfrak{V}$ is isomorphic to the quotient $\mathfrak{A}/M$, where M is the ideal that is the dual of the special open set $\sim V$. Inversely, if M is an ideal in $\mathfrak{A}$, then the dual space of* *the quotient algebra $\mathfrak{A}/M$ is homeo-isomorphic to the inner subspace that is the restriction of $\mathfrak{U}$ to V, where V is the complement of the special open set that is the dual of M.*

Proof. Each ideal M in $\mathfrak{A}$ uniquely determines an inner subspace of $\mathfrak{U}$, by Theorems 2.23 and 2.27, namely the inner subspace whose universe is the special closed set that is the complement of the special open set that is the dual of M. Conversely, each inner subspace $\mathfrak{V}$ of $\mathfrak{U}$ uniquely determines an ideal in $\mathfrak{A}$, namely the ideal that is the dual of the special open set that is the complement of the universe of $\mathfrak{V}$. The correspondence mapping the inner subspace $\mathfrak{V}$ to the quotient $\mathfrak{A}/M$ is therefore a bijection from the set of inner subspaces of $\mathfrak{U}$ to the set of quotients of $\mathfrak{A}$, by Theorem 2.23.

To prove the second assertion of the theorem, suppose $\mathfrak{V}$ is an inner subspace of the relational space $\mathfrak{U}$, and let $\mathfrak{B}$ be the dual algebra of $\mathfrak{V}$. Since $\mathfrak{V}$ is, in particular, an inner substructure of $\mathfrak{U}$, the identity function ϑ on $\mathfrak{V}$ is a bounded monomorphism from $\mathfrak{V}$ into $\mathfrak{U}$, by Corollary 1.14. Also, ϑ is continuous, by the definition of the inherited topology on $\mathfrak{V}$. Indeed, if H is an open set in $\mathfrak{U}$, then

$$\vartheta^{-1}(H) = H \cap V,$$

which is an open set in $\mathfrak{V}$, by the definition of the inherited topology; so the inverse image under ϑ of every open set in $\mathfrak{V}$ is an open set in $\mathfrak{U}$.

The dual of ϑ is the epimorphism φ from $\mathfrak{A}$ to $\mathfrak{B}$ that is defined by

$$\varphi(F) = \vartheta^{-1}(F) = \{u \in V : \vartheta(u) \in F\} = F \cap V$$

for elements F in $\mathfrak{A}$, that is to say, for clopen subsets F of $\mathfrak{U}$, by Theorem 2.16 and the assumption that ϑ is the identity function. The kernel of φ is the set of elements mapped to the empty set by φ. Since

$$\varphi(F) = \varnothing \qquad \text{if and only if} \qquad F \cap V = \varnothing,$$
$$\text{if and only if} \qquad F \subseteq {\sim}V,$$

the kernel of φ is just the ideal of clopen subsets of ${\sim}V$. In other words, the kernel is the ideal M that is the dual of the special open set ${\sim}V$. Because φ is an epimorphism from $\mathfrak{A}$ to $\mathfrak{B}$ with kernel M, the quotient $\mathfrak{A}/M$ is isomorphic to $\mathfrak{B}$ via the function that maps each coset F/M to the intersection $\varphi(F) = F \cap V$, by the First Isomorphism Theorem for Boolean algebras with operators.

To prove the third assertion of the theorem, consider an arbitrary ideal M in $\mathfrak{A}$. The dual of M is the special open set that is the union of the sets in M. If V is the complement of this special open set, and if $\mathfrak{V}$ is the restriction of $\mathfrak{U}$ to V, then $\mathfrak{V}$ is an inner subspace of $\mathfrak{U}$, by Theorem 2.27. Let $\mathfrak{B}$ be the dual algebra of $\mathfrak{V}$. The dual ideal of the special open set $\sim V$ is M, by Theorem 2.23, so $\mathfrak{B}$ is isomorphic to the quotient $\mathfrak{A}/M$, by the observations of the preceding paragraphs. Apply Theorem 2.18 to conclude that the dual space of the quotient $\mathfrak{A}/M$ is homeo-isomorphic to the dual space of $\mathfrak{B}$. Since $\mathfrak{B}$ is the dual algebra of $\mathfrak{V}$, the dual space of $\mathfrak{B}$ is, by definition, the second dual of $\mathfrak{V}$, and $\mathfrak{V}$ is homeo-isomorphic to its second dual, by Theorem 2.11. It follows that the dual space of the quotient $\mathfrak{A}/M$ is homeo-isomorphic to $\mathfrak{V}$, as claimed.   □

For some remarks concerning the relationship of Theorem 2.29 to the duality between inner subspaces and homomorphic images that is known from the literature, see the end of Section 2.8.

2.7 Duality for Subuniverses

The other half of the sub-quotient duality involves a duality between subuniverses of Boolean algebras with operators and special congruences on relational spaces. There is a corresponding duality between subalgebras of Boolean algebras with operators and quotients of relational spaces.

Recall that if U is a topological space, and Θ an equivalence relation on U, then the set V of equivalence classes of Θ can be turned into a topological space by declaring a subset F of V to be open just in case the union of the equivalence classes in F is an open subset of U. The set V endowed with this *quotient topology* is called the *quotient space* of U modulo Θ. A quotient of a compact space is necessarily compact, but a quotient of a Hausdorff space or a Boolean space need not be Hausdorff or Boolean. An equivalence relation Θ on U is said to be *Boolean* if for any two elements in $\mathfrak{U}$ that are inequivalent modulo Θ, there is a clopen subset of $\mathfrak{U}$ that is compatible with Θ (that is to say, it is a union of equivalence classes of Θ—see the remarks preceding Lemma 1.21) and that contains one of the two elements but not the other. It is not difficult to check that for any equivalence relation Θ on a Boolean space, the quotient space is Boolean if and only if Θ is a Boolean relation (see, for example, Lemma 1 on p. 362 of [10]). The quotient function that maps each element in a Boolean space U to its equivalence class modulo a Boolean relation Θ is a continuous function from U onto the quotient space V, by the definition of the quotient topology. Conversely, if ϑ is a continuous mapping from a Boolean space U onto a Boolean space W, then the *kernel* of ϑ, that is to say, the relation Θ defined by

$$r \equiv s \mod \Theta \quad \text{if and only if} \quad \vartheta(r) = \vartheta(s),$$

is a Boolean relation on U, and the quotient of U modulo Θ is homeomorphic to W via the mapping that take each equivalence class u/Θ to the element $\vartheta(u)$.

To defined the appropriate notion of a congruence on a relational space, we must combine the notions of a Boolean relation and a bounded congruence (see Definition 1.20).

Definition 2.30. A binary relation Θ on a relational space $\mathfrak{U}$ is called a *relational congruence* if Θ is a bounded congruence on $\mathfrak{U}$ and simultaneously a Boolean relation with respect to the topology on $\mathfrak{U}$. $\square$

Every relational congruence on a relational space gives rise to a subuniverse of the dual algebra.

Lemma 2.31. *If Θ is a relational congruence on a relational space $\mathfrak{U}$, then the set of clopen subsets of $\mathfrak{U}$ that are compatible with Θ is a subuniverse of the dual algebra of $\mathfrak{U}$.*

Proof. Let $\mathfrak{A}$ be the dual algebra of $\mathfrak{U}$, and B the set of clopen subsets of $\mathfrak{U}$ that are compatible with Θ. It is to be shown that B is a subuniverse of $\mathfrak{A}$. Clearly, the empty set is clopen and compatible with Θ, so it belongs to B. Also, the union of two clopen sets is clopen, and the complement of a clopen set is clopen, in any topological space; and the union of two sets compatible with Θ is compatible with Θ, and the complement of a set compatible with Θ is compatible with Θ. Consequently, B is closed under the Boolean operations of $\mathfrak{A}$.

To show that B is closed under the operators of $\mathfrak{A}$, consider the case of a binary operator $\circ$ that is defined in terms of a ternary relation R in $\mathfrak{U}$. Let F and G be sets in B. The product of $F \circ G$ in $\mathfrak{A}$, which is the image set defined by

$$F \circ G = R^*(F \times G) = \{w : R(u, v, w) \text{ for some } u \in F \text{ and } v \in G\},$$

is clopen in $\mathfrak{U}$, by the assumption that $\mathfrak{U}$ is a relational space (see Definition 2.2), and it is compatible with Θ because Θ is assumed to be a bounded congruence on $\mathfrak{U}$ (see the argument in the corresponding part of the proof of Lemma 1.21). Consequently, $F \circ G$ belongs to B, so B is in fact a subuniverse of $\mathfrak{A}$. □

There is a type of converse to the preceding lemma that is also true.

Lemma 2.32. *If B is a subuniverse of the dual algebra of a relational space $\mathfrak{U}$, then the relation Θ on $\mathfrak{U}$ that is defined by $r \equiv s \mod \Theta$ if and only if r and s belong to the same sets in B is a relational congruence on $\mathfrak{U}$.*

Proof. We give an indirect proof that makes use of the epi-mono duality. Suppose $\mathfrak{A}$ is the dual algebra of $\mathfrak{U}$, and $\mathfrak{V}$ the dual space of $\mathfrak{A}$. Thus, $\mathfrak{V}$ is the second dual of $\mathfrak{U}$: the elements in $\mathfrak{V}$ are the sets of the form

$$X_r = \{F \in A : r \in F\}, \tag{1}$$

and the function ϱ that maps each element r in $\mathfrak{U}$ to the set X_r is a homeo-isomorphism from $\mathfrak{U}$ to $\mathfrak{V}$, by Theorem 2.11.

Write $\mathfrak{B}$ for the subalgebra of $\mathfrak{A}$ with universe B. The identity function φ on $\mathfrak{B}$ is a monomorphism from $\mathfrak{B}$ into $\mathfrak{A}$, since $\mathfrak{B}$ is a subalgebra of $\mathfrak{A}$. The dual of φ is the function ϑ on $\mathfrak{V}$ determined by

$$\vartheta(X_r) = \varphi^{-1}(X_r) = \{F \in B : \varphi(F) \in X_r\}$$
$$= \{F \in B : F \in X_r\} = \{F \in B : r \in F\}, \tag{2}$$

by the definition of ϑ, the definition of the inverse image of a set under the function φ, the assumption that φ is the identity function on $\mathfrak{B}$, and (1). This dual is a continuous bounded epimorphism from $\mathfrak{V}$ to the dual relational space of $\mathfrak{B}$, by Theorem 2.18. The kernel of ϑ is, by definition, the relation Ψ on $\mathfrak{V}$ that is defined by

$$X_r \equiv X_s \mod \Psi \qquad \text{if and only if} \qquad \vartheta(X_r) = \vartheta(X_s). \tag{3}$$

Since the kernel of a bounded homomorphism is a bounded congruence, by Lemma 1.27, and the kernel of a continuous mapping between Boolean spaces is a Boolean relation (see the remark before Definition 2.30), it follows that Ψ is a relational congruence on $\mathfrak{U}$, by Definition 2.30.

The equations in (2) imply that

$$\vartheta(X_r) = \vartheta(X_s) \quad \text{if and only if}$$
$$\{F \in B : r \in F\} = \{F \in B : s \in F\}.$$

The right side of this equivalence just says that r and s belong to the same sets in B, which is exactly what it means for the elements r and s to be equivalent modulo Θ. Combine these observations with (3) to conclude that

$$X_r \equiv X_s \quad \text{mod } \Psi \qquad \text{if and only if} \qquad r \equiv s \quad \text{mod } \Theta. \qquad (4)$$

The equivalence in (4) implies that the relation Θ is the inverse image, under the homeo-isomorphism ϱ, of the relation Ψ. Homeo-isomorphisms preserve all algebraic and topological properties. In particular, since Ψ is a relational congruence on $\mathfrak{V}$, it follows that Θ must be a relational congruence on $\mathfrak{U}$. $\square$

We continue with the assumption that $\mathfrak{U}$ is a relational space, and $\mathfrak{A}$ its dual algebra. If Θ is a relational congruence on $\mathfrak{U}$, then the subuniverse of $\mathfrak{A}$ consisting of the clopen sets that are compatible with Θ (Lemma 2.31) is called the (*first*) *dual*, or the *dual subuniverse*, of Θ. Similarly, if B is a subuniverse of $\mathfrak{A}$, then the relational congruence on $\mathfrak{U}$ consisting of the pairs of elements that belong to precisely the same sets in B (Lemma 2.32) is called the (*first*) *dual*, or the *dual relational congruence*, of B.

Each relational congruence on $\mathfrak{U}$ is its own second dual, that is to say, if Θ is a relational congruence on $\mathfrak{U}$ with dual subuniverse B, and if Ψ is the dual relational congruence of B, then $\Theta = \Psi$. The proof depends on two facts. First, the sets in B are all compatible with Θ, by the definition of B. In particular, if r and s are elements in $\mathfrak{U}$ that are congruent modulo Θ, then for any set F in B, either r and s are both in F, or neither r nor s is in F, by the compatibility of F with Θ. It follows that r and s belong to the same sets in B, and therefore r and s are congruent modulo Ψ, by the definition of Ψ as the dual of the subuniverse B. Second, Θ is a Boolean relation, so if elements r and s are not congruent modulo Θ, then there must be a clopen subset F of $\mathfrak{U}$ that is compatible with Θ and that contains r but not s. The set F belongs to B, by the definition of B, so the elements r and s do not belong to precisely the same sets in B and consequently r and s are not equivalent modulo Ψ. Conclusion: $\Theta = \Psi$.

Each subuniverse of $\mathfrak{A}$ is also equal to its second dual. For the proof, suppose B is a subuniverse of $\mathfrak{A}$ with dual relational congruence Θ, and let C be the dual subuniverse of Θ. It is to be shown that $B = C$. The proof requires a number of preliminary observations. Write

$$B/\Theta = \{F/\Theta : F \in B\} \qquad \text{and} \qquad C/\Theta = \{F/\Theta : F \in C\},$$

where
$$F/\Theta = \{r/\Theta : r \in F\}.$$

Since Θ is, in particular, a Boolean relation on the topological space U, the quotient space U/Θ is a Boolean space, and the clopen subsets of this quotient space are the quotients of the clopen subsets of U that are compatible with Θ (see Lemma 1 on p. 362 of [10]). By definition, C consists of all clopen subsets of U that are compatible with Θ, so C/Θ is the set of all clopen subsets of U/Θ.

Turn now to the quotient B/Θ. Every set in B is a clopen subset of U, because B is a subuniverse of the algebra $\mathfrak{A}$ of all clopen subsets of U. Furthermore, every set in B is compatible with Θ. Indeed, if r and s are elements in $\mathfrak{U}$ that are equivalent modulo Θ, then r and s belong to exactly the same sets in B, by the definition of Θ as the dual of the subuniverse B. Consequently, if F is a set in B, and if r is in F, then s is in F. The presence of an element r in F therefore implies that the entire equivalence class r/Θ is included in F, so F is compatible with Θ. It follows from these observations that the set B is included in C, and therefore the quotient B/Θ is included in C/Θ.

The next step is to show that B/Θ is closed under the operations of union and complement. To this end, suppose that F and G are sets in B. The union $F \cup G$ and the complement $\sim F$ are also in B, since B is a subuniverse of $\mathfrak{A}$. To show that B/Θ is closed under union, it therefore suffices to check that

$$(F/\Theta) \cup (G/\Theta) = (F \cup G)/\Theta.$$

For an arbitrary element r in U, we have

$$r/\Theta \in (F/\Theta) \cup (G/\Theta) \quad \text{if and only if} \quad r/\Theta \in F/\Theta \text{ or } r/\Theta \in G/\Theta,$$
$$\text{if and only if} \quad r \in F \text{ or } r \in G,$$
$$\text{if and only if} \quad r \in F \cup G,$$
$$\text{if and only if} \quad r/\Theta \in (F \cup G)/\Theta,$$

by the definition of the union of two sets and by the compatibility of
the sets F, G, and $F \cup G$ with the congruence Θ. Consequently, the
desired equality does hold. A similar argument shows that

$$\sim(F/\Theta) = (\sim F)/\Theta$$

(where the first complement is formed in B/Θ, and the second in B),
so that B/Θ is closed under complement.

The third step is to observe that the sets in B/Θ—which are clopen
subsets of the quotient space U/Θ—separate points in the quotient
space. Indeed, if r/Θ and s/Θ are distinct points in the quotient space,
then the definition of the relation Θ implies that there must be a set F
in B that contains the element r but not the element s. The clopen
set F/Θ belongs to B/Θ, and it contains the point r/Θ but not the
point s/Θ (because F is compatible with Θ), so F/Θ separates these
two points.

It has been shown that B/Θ is a Boolean algebra of clopen subsets
of U/Θ that separates points in U/Θ. Any Boolean algebra of clopen
sets that separates points in a Boolean space is necessarily the Boolean
algebra of all clopen sets in the space (see Lemma 1 on p. 305 of [10]).
Consequently, B/Θ is the Boolean algebra of all clopen sets in U/Θ,
that is to say,

$$B/\Theta = C/\Theta.$$

Each set F that belongs to B or to C is the union of the equivalence
classes in the quotient F/Θ, because F is compatible with Θ. The
equality at the end of the last paragraph therefore implies that $B = C$.
In more detail, if F is in C, then the quotient set F/Θ belongs to C/Θ,
and therefore also to B/Θ, by the equality at the end of the last
paragraph. Consequently, $F/\Theta = G/\Theta$ for some set G in B. Since

$$F = \bigcup(F/\Theta) = \bigcup(G/\Theta) = G,$$

it follows that F belongs to B. Thus, C is included in B. The reverse
inclusion was established above. This completes the proof that B is its
own second dual.

Consider now two arbitrary relational congruences on $\mathfrak{U}$, say Θ
and Ψ, with dual subuniverses B and C respectively. It is not difficult to check that

$$\Theta \subseteq \Psi \qquad \text{if and only if} \qquad C \subseteq B.$$

For the proof, assume first that C is included in B. The relations Θ and Ψ are their own second duals, so they are the first duals of the subuniverses B and C respectively. If r and s are elements in $\mathfrak{U}$ that are equivalent modulo Θ, then these two elements belong to the same sets in B, by definition of the first dual of B. Since C is assumed to be included in B, it follows that r and s belong to the same sets in C, and therefore r and s are equivalent modulo Ψ, by the definition of first dual of C. Thus Θ is included in Ψ.

To prove the reverse implication, assume that Θ is included in Ψ. The subuniverses B and C are defined to be the sets of clopen subsets of $\mathfrak{U}$ that are compatible with Θ and Ψ respectively. Consequently, if F is an arbitrary element in C, then F must be clopen and compatible with Ψ. It follows from the assumed inclusion that F must also be compatible with Θ, and therefore must belong to B. In more detail, if r is an element in F, then the entire equivalence class r/Ψ is included in F, by the compatibility of F with Ψ. Since Θ is included in Ψ, the equivalence class r/Θ is included in the equivalence class r/Ψ, and therefore r/Θ is also included in F. Thus, F is also compatible with Θ, and therefore F belongs to B, as claimed. This argument shows that C is included in B.

The preceding arguments imply that the function ϱ mapping each relational congruence on $\mathfrak{U}$ to its dual subuniverse of $\mathfrak{A}$ is a dual lattice isomorphism. Indeed, if δ is the function mapping each subuniverse of $\mathfrak{A}$ to its dual relational congruence on $\mathfrak{U}$, then

$$(\delta \circ \varrho)(\Theta) = \Theta \qquad \text{and} \qquad (\varrho \circ \delta)(B) = B$$

for each relational congruence Θ and each subuniverse B, because Θ is the dual of the dual of Θ, and B is the dual of the dual of B. Thus, $\delta \circ \varrho$ is the identity function on the lattice of relational congruences on $\mathfrak{U}$, and $\varrho \circ \delta$ is the identity function on the lattice of subuniverses of $\mathfrak{A}$, so ϱ and δ are bijections and inverses of one another. Since a relational congruence Θ is included in a relational congruence Ψ if and only if the dual of Ψ is included in the dual of Θ, that is to say, if and only if $\varrho(\Psi)$ is included in $\varrho(\Theta)$, the bijection ϱ reverses the lattice partial ordering, and is therefore a dual lattice isomorphism. The results of this section are summarized in the following theorem.

Theorem 2.33. *Suppose $\mathfrak{U}$ is a relational space and $\mathfrak{A}$ its dual algebra. The dual of every relational congruence on $\mathfrak{U}$ is a subuniverse of $\mathfrak{A}$, and the dual of every subuniverse of $\mathfrak{A}$ is a relational congruence on $\mathfrak{U}$.*

The second dual of every relational congruence and of every subuniverse is itself. The function mapping each relational congruence on $\mathfrak{U}$ to its dual subuniverse of $\mathfrak{A}$ is a dual lattice isomorphism from the lattice of relational congruences on $\mathfrak{U}$ to the lattice of subuniverses of $\mathfrak{A}$.

A related result, discovered independently by Celani, is given in Theorem 27 of [2].

2.8 Duality for Subalgebras

The duality between relational congruences and subuniverses implies a corresponding duality between quotient spaces and subalgebras. By the quotient of a relational space, we mean the following.

Definition 2.34. The *quotient* of a relational space $\mathfrak{U}$ modulo a relational congruence Θ is the quotient relational structure of $\mathfrak{U}$ (see Definition 1.26) endowed with the quotient topology. The quotient is denoted by $\mathfrak{U}/\Theta$. The *quotient function*, or *quotient mapping*, from $\mathfrak{U}$ to $\mathfrak{U}/\Theta$ is the function that maps each element u in $\mathfrak{U}$ to its congruence class u/Θ. $\square$

The first observation to make is that every continuous bounded homomorphism gives rise to a relational congruence, namely its kernel.

Lemma 2.35. *If ϑ is a continuous bounded homomorphism from a relational space $\mathfrak{U}$ to a relational space $\mathfrak{V}$, then the kernel of ϑ is a relational congruence on $\mathfrak{U}$.*

Proof. The kernel of a continuous mapping between Boolean spaces is a Boolean relation, so the kernel of ϑ—call it Θ—is a Boolean relation. The kernel of a bounded homomorphism is a bounded congruence, by Lemma 1.27, so Θ is a bounded congruence on $\mathfrak{U}$. Therefore, Θ is a relational congruence, by Definition 2.30. $\square$

The converse of the preceding lemma says that every relational congruence gives rise to a continuous bounded epimorphism. This observation is the analogue of Theorem 1.28, and its proof presents no difficulties.

Theorem 2.36. *If Θ is a relational congruence on a relational space $\mathfrak{U}$, then the corresponding quotient mapping is a continuous bounded epimorphism from $\mathfrak{U}$ to the quotient $\mathfrak{U}/\Theta$ that has Θ as its kernel.*

Proof. The quotient mapping ϑ is a bounded epimorphism from $\mathfrak{U}$ to $\mathfrak{U}/\Theta$ with kernel Θ, by Theorem 1.28, and ϑ is continuous because $\mathfrak{U}/\Theta$ is endowed with the quotient topology. $\square$

The proof that quotients of relational spaces are relational spaces is more involved.

Lemma 2.37. *The quotient of a relational space modulo a relational congruence is a relational space.*

Proof. Suppose Θ is a relational congruence on a relational space $\mathfrak{U}$, and ϑ the quotient mapping from $\mathfrak{U}$ onto $\mathfrak{U}/\Theta$. The quotient $\mathfrak{U}/\Theta$ is certainly a relational structure. The congruence Θ is, by definition, a Boolean relation, so the quotient topology on the quotient of the universe of $\mathfrak{U}$ turns that quotient into a Boolean space and turns ϑ into a continuous function. It remains to check that the fundamental relations in the quotient are clopen, and those of rank at least two are continuous, in the quotient topology.

Focus on the case of a ternary relation R. To see that R is clopen in $\mathfrak{U}/\Theta$, consider clopen subsets F and G of $\mathfrak{U}/\Theta$. It is to be shown that the image set

$$R^*(F \times G) = \{t \in U/\Theta : R(r, s, t) \text{ for some } r \in F \text{ and } s \in G\} \quad (1)$$

is clopen in the quotient topology. The inverse images

$$\bar{F} = \vartheta^{-1}(F) \qquad \text{and} \qquad \bar{G} = \vartheta^{-1}(G) \tag{2}$$

are clopen subsets of $\mathfrak{U}$ because the mapping ϑ is continuous, and they are compatible with Θ because ϑ is the quotient mapping, so the image set

$$R^*(\bar{F} \times \bar{G}) = \{w \in U : R(u, v, w) \text{ for some } u \in \bar{F} \text{ and } v \in \bar{G}\} \quad (3)$$

is certainly clopen in $\mathfrak{U}$, by the assumption that $\mathfrak{U}$ is a relational space (see Definition 2.2). Furthermore, the image set in (3) is compatible with Θ. Indeed, consider any element w in the image set, and suppose that $\bar{w}$ belongs to the congruence class w/Θ; it must be shown that $\bar{w}$ is also in the image set. Since w is in (3), there are elements u in $\bar{F}$ and v

in $\bar{G}$ such that $R(u, v, w)$ holds in $\mathfrak{U}$. The congruence Θ is bounded, by definition, and $\bar{w} \equiv w \mod \Theta$, by assumption, so there must be elements $\bar{u}$ and $\bar{v}$ in $\mathfrak{U}$ such that

$$\bar{u} \equiv u \mod \Theta, \qquad \bar{v} \equiv v \mod \Theta, \qquad \text{and} \qquad R(\bar{u}, \bar{v}, \bar{w}). \qquad (4)$$

The sets $\bar{F}$ and $\bar{G}$ are compatible with Θ and contain the elements u and v respectively, so they also contain the elements $\bar{u}$ and $\bar{v}$ respectively, by the definition of compatibility. It follows that $\bar{w}$ belongs to the image set in (3), by the final part of (4). Thus, the image set in (3) is compatible with Θ, as claimed. The quotient of a clopen set in $\mathfrak{U}$ that is compatible with Θ is a clopen set in the quotient $\mathfrak{U}/\Theta$ (see the remarks following Lemma 2.32). Consequently, the quotient set

$$R^*(\bar{F} \times \bar{G})/\Theta = \{w/\Theta : w \in R^*(\bar{F} \times \bar{G})\}$$

is clopen in $\mathfrak{U}/\Theta$.

In view of the preceding observations, to prove that the set in (1) is clopen it suffices to show that

$$R^*(F \times G) = R^*(\bar{F} \times \bar{G})/\Theta. \qquad (5)$$

Consider an arbitrary element t in $\mathfrak{U}/\Theta$, and assume first that t belongs to the left side of (5). There are then elements r in F and s in G such that $R(r, s, t)$ holds in $\mathfrak{U}/\Theta$, by (1). The relation R in $\mathfrak{U}/\Theta$ is defined to be the quotient of the corresponding relation in $\mathfrak{U}$, so there must be elements u, v, and w in $\mathfrak{U}$ such that

$$r = u/\Theta, \qquad s = v/\Theta, \qquad t = w/\Theta, \qquad \text{and} \qquad R(u, v, w)$$

in $\mathfrak{U}$. The function ϑ maps each element in $\mathfrak{U}$ to its quotient, so we must have

$$\vartheta(u) = u/\Theta = r, \qquad \vartheta(v) = v/\Theta = s, \qquad \vartheta(w) = w/\Theta = t. \qquad (6)$$

Since r and s belong to the sets F and G respectively, it follows from (6) and (2) that u and v belong to the inverse image sets $\bar{F}$ and $\bar{G}$ respectively. Consequently, w belongs to the set in (3), and therefore the quotient w/Θ, which is equal to t, belongs to the quotient on the right side of (5).

Assume now that t belongs to the right side of (5). In this case, t is equal to w/Θ for some element w in $R^*(\bar{F} \times \bar{G})$, so there must be elements u in $\bar{F}$ and v in $\bar{G}$ such that $R(u, v, w)$ holds in $\mathfrak{U}$, by (3).

Write $r = u/\Theta$ and $s = v/\Theta$, and observe two things: first, $R(r, s, t)$ must hold in the quotient $\mathfrak{U}/\Theta$, by the definition of the relation R in the quotient; and second, r and s must belong to the sets F and G respectively, by (2), (6), and the fact that u and v are in $\bar{F}$ and $\bar{G}$ respectively. Therefore, t belongs to the left side of (5), by (1). Conclusion: the equation in (5) holds, so the relation R in $\mathfrak{U}/\Theta$ is clopen, as claimed.

The final task is to prove that the relation R in $\mathfrak{U}/\Theta$ is continuous. In view of Theorem 2.14, it suffices to prove that R is weakly continuous in the sense of Definition 2.12. Consider an element t in $\mathfrak{U}/\Theta$, and write

$$H = \{(r, s) \in U/\Theta \times U/\Theta : R(r, s, t)\}. \tag{7}$$

It is to be shown that the set H is closed in the product topology on the space

$$U/\Theta \times U/\Theta. \tag{8}$$

The element t is, by assumption, the quotient of some element in $\mathfrak{U}$, say

$$\vartheta(w) = w/\Theta = t. \tag{9}$$

The set

$$\bar{H} = \{(u, v) \in U \times U : R(u, v, w)\} \tag{10}$$

is certainly closed in $\mathfrak{U}$, because $\mathfrak{U}$ is a relational space and therefore the relation R in $\mathfrak{U}$ is weakly continuous, by Theorem 2.13. The quotient function ϑ is continuous, because U/Θ is endowed with the quotient topology, so the induced function $\bar{\vartheta}$ defined by

$$\bar{\vartheta}(u, v) = (\vartheta(u), \vartheta(v))$$

for u and v in U is a continuous mapping from $U \times U$ to $U/\Theta \times U/\Theta$. A continuous function from a compact space to a Hausdorff space maps closed sets to closed sets (see Corollary 1 on p. 315 of [10]). Since the spaces $U \times U$ and $U/\Theta \times U/\Theta$ are both Boolean (and hence both compact Hausdorff spaces), and since the set $\bar{H}$ is closed, the image set

$$\bar{\vartheta}(\bar{H}) = \{\bar{\vartheta}(u, v) : (u, v) \in \bar{H}\}$$

must also be closed.

In view of the observations of the preceding paragraph, to prove that the set H in (7) is closed in the space (8), it suffices to show that

$$\bar{\vartheta}(\bar{H}) = H. \tag{11}$$

To this end, consider an arbitrary pair (u, v) in $\bar{H}$. Because ϑ is a homomorphism, by Theorem 2.36, and because $R(u, v, w)$ holds in $\mathfrak{U}$, by (10), we have $R(\vartheta(u), \vartheta(v), \vartheta(w))$ in $\mathfrak{U}/\Theta$. Also, $\vartheta(w) = t$, by (9), so it follows from (7) (with r and s replaced by $\vartheta(u)$ and $\vartheta(v)$ respectively) that the pair $(\vartheta(u), \vartheta(v))$ belongs to the set H. Apply the definition of the function $\bar{\vartheta}$ to see that $\bar{\vartheta}(u, v)$ must belong to H. Conclusion: the image set $\bar{\vartheta}(\bar{H})$ is included in H.

To establish the reverse inclusion, assume that (r, s) is an arbitrary pair in H. In this case, $R(r, s, t)$ holds in the quotient $\mathfrak{U}/\Theta$, by (7), and therefore $R(r, s, \vartheta(w))$ holds in the quotient, by (9). The quotient mapping ϑ is a bounded epimorphism, by Theorem 2.36, so there must be elements u and v in $\mathfrak{U}$ such that

$$\vartheta(u) = r, \qquad \vartheta(v) = s, \qquad \text{and} \qquad R(u, v, w).$$

The pair (u, v) belongs to $\bar{H}$, by (10), and

$$\bar{\vartheta}(u, v) = (\vartheta(u), \vartheta(v)) = (r, s),$$

so (r, s) belongs to the image set $\bar{\vartheta}(\bar{H})$. Conclusion: H is included in the set $\bar{\vartheta}(\bar{H})$. This completes the proof of (11), so H is closed in the product space (8), as claimed. Consequently, the relation R in $\mathfrak{U}/\Theta$ is weakly continuous, and therefore continuous.   □

The next theorem is the analogue for relational spaces of the First Isomorphism Theorem for algebras. It says that up to homeo-isomorphisms, the only continuous bounded homomorphic images of a relational space are the relational quotients of that space.

Theorem 2.38. *Every continuous bounded homomorphic image of a relational space $\mathfrak{U}$ is homeo-isomorphic to a quotient of $\mathfrak{U}$ modulo a relational congruence on $\mathfrak{U}$. In fact, if ϑ is a continuous bounded epimorphism from $\mathfrak{U}$ to a relational space $\mathfrak{V}$, and if Θ is the kernel of ϑ, then the function mapping u/Θ to $\vartheta(u)$ for each u in $\mathfrak{U}$ is a homeo-isomorphism from $\mathfrak{U}/\Theta$ to $\mathfrak{V}$.*

Proof. The kernel Θ is a relational congruence on $\mathfrak{U}$, by Lemma 2.35. The function that maps each congruence class u/Θ to the element $\vartheta(u)$ is a well-defined homeomorphism from the topological structure of $\mathfrak{U}/\Theta$ to the topological structure of $\mathfrak{V}$, by the First Isomorphism Theorem for topological spaces, and it is an isomorphism from the algebraic structure of $\mathfrak{U}/\Theta$ to the algebraic structure of $\mathfrak{V}$, by First Isomorphism Theorem for relational structures (see Theorem 1.29). Consequently, this function is a homeo-isomorphism from $\mathfrak{U}/\Theta$ to $\mathfrak{V}$.   □

We are now ready to establish the duality between quotient relational spaces and subalgebras of relational algebras.

Theorem 2.39. *There is a bijective correspondence between the quotients of a relational space $\mathfrak{U}$ and the subalgebras of its dual algebra $\mathfrak{A}$. If Θ is a relational congruence on $\mathfrak{U}$, then the dual algebra of the quotient $\mathfrak{U}/\Theta$ is isomorphic to the subalgebra of $\mathfrak{A}$ whose universe is the dual subuniverse of Θ. Inversely, if $\mathfrak{B}$ is a subalgebra of $\mathfrak{A}$, then the dual space of $\mathfrak{B}$ is homeo-isomorphic to the quotient space $\mathfrak{U}/\Theta$, where Θ is the dual relational congruence of the universe of $\mathfrak{B}$.*

Proof. Each relational congruence Θ on $\mathfrak{U}$ uniquely determines a subalgebra of the dual algebra $\mathfrak{A}$, namely the subalgebra whose universe is the dual subuniverse of Θ, that is to say, the subalgebra whose universe is the set of clopen subsets of $\mathfrak{U}$ that are compatible with Θ, by Lemma 2.31. Conversely, each subalgebra $\mathfrak{B}$ of $\mathfrak{A}$ uniquely determines a relational congruence on $\mathfrak{U}$, namely the relational congruence that is the dual of the universe of $\mathfrak{B}$, that is to say, the congruence consisting of the pairs of elements from $\mathfrak{U}$ that belong to exactly the same sets in $\mathfrak{B}$, by Lemma 2.32. Because subuniverses and relational congruences are their own second duals, it follows that the correspondence mapping each quotient space $\mathfrak{U}/\Theta$ to the dual subalgebra $\mathfrak{B}$ is a bijection from the set of quotient spaces of $\mathfrak{U}$ to the set of subalgebras of $\mathfrak{A}$, by Theorem 2.33.

To prove the second assertion of the theorem, suppose Θ is a relational congruence on $\mathfrak{U}$, and let $\mathfrak{C}$ be the dual algebra of the quotient space $\mathfrak{U}/\Theta$. The elements in $\mathfrak{C}$ are the clopen subsets of $\mathfrak{U}/\Theta$, so they are the sets of the form

$$F/\Theta = \{u/\Theta : u \in F\},$$

where F ranges over the clopen subsets of $\mathfrak{U}$ that are compatible with Θ. The quotient function ϑ from $\mathfrak{U}$ to $\mathfrak{U}/\Theta$, which maps each element u to its congruence class u/Θ, is a continuous bounded epimorphism, by Theorem 2.36. The dual of ϑ is the monomorphism φ from $\mathfrak{C}$ into $\mathfrak{A}$ that is defined by

$$\varphi(F/\Theta) = \vartheta^{-1}(F/\Theta)$$

for each element F/Θ in $\mathfrak{C}$, by Theorem 2.16. For clopen sets F that are compatible with Θ, the inverse image under ϑ of the quotient set F/Θ is just F, because F includes the equivalence class of each of its elements. Consequently,

$$\varphi(F/\Theta) = F$$

for each set F/Θ in $\mathfrak{C}$. It follows from these observations that φ maps the universe of $\mathfrak{C}$ bijectively to the subuniverse of $\mathfrak{A}$ consisting of the clopen subsets of $\mathfrak{U}$ that are compatible with Θ, so φ is an isomorphism from $\mathfrak{C}$ to the subalgebra of $\mathfrak{A}$ whose universe is the dual of Θ.

To prove the third assertion of the theorem, consider an arbitrary subalgebra $\mathfrak{B}$ of $\mathfrak{A}$, and let Θ be the relational congruence on $\mathfrak{U}$ that is the dual of the universe of $\mathfrak{B}$. The dual of Θ is, by definition, the second dual of the universe of $\mathfrak{B}$, and the universe of $\mathfrak{B}$ is its own second dual, by Theorem 2.33. Apply the part of the theorem already proved to conclude that the dual algebra of $\mathfrak{U}/\Theta$—call it $\mathfrak{C}$—is isomorphic to $\mathfrak{B}$. The dual relational spaces of these two algebras must therefore be homeo-isomorphic, by Theorem 2.18. The dual relational space of $\mathfrak{C}$ is, by definition, the second dual of the quotient space $\mathfrak{U}/\Theta$, and these two spaces are homeo-isomorphic, by Theorem 2.11. Consequently, $\mathfrak{U}/\Theta$ is homeo-isomorphic to the dual space of $\mathfrak{B}$, as desired. $\square$

As in the case of relational structures and their dual complete and atomic Boolean algebras with operators (see the remarks at the end of Section 1.9), there are weaker, less explicit versions of Theorems 2.29 and 2.39 that are known from the literature and that follow almost immediately from the epi-mono duality between continuous bounded homomorphisms and homomorphisms formulated in Theorem 2.20. To describe these weaker versions, consider two relational spaces $\mathfrak{U}$ and $\mathfrak{V}$ with dual algebras $\mathfrak{A}$ and $\mathfrak{B}$ respectively.

If $\mathfrak{V}$ is homeo-isomorphic to an inner subspace of $\mathfrak{U}$, say via a continuous bounded monomorphism ϑ, then the dual of ϑ is an epimorphism φ from $\mathfrak{A}$ to $\mathfrak{B}$, so that $\mathfrak{B}$ is a homomorphic image of $\mathfrak{A}$. Conversely, if $\mathfrak{B}$ is a homomorphic image of $\mathfrak{A}$, say via an epimorphism φ, then the dual of φ is a continuous bounded monomorphism ϑ from $\mathfrak{V}$ to $\mathfrak{U}$, so that $\mathfrak{V}$ is isomorphic to an inner subspace of $\mathfrak{U}$. Thus, there is a duality between (homeo-isomorphic copies of) inner subspaces of a relational space $\mathfrak{U}$ and homomorphic images of the dual algebra $\mathfrak{A}$.

In a similar vein, if $\mathfrak{V}$ is a continuous bounded homomorphic image of $\mathfrak{U}$, say via a continuous bounded epimorphism ϑ, then the dual of ϑ is a monomorphism φ from $\mathfrak{B}$ to $\mathfrak{A}$, so that $\mathfrak{B}$ is isomorphic to a subalgebra of $\mathfrak{A}$. Conversely, if $\mathfrak{B}$ is isomorphic to a subalgebra of $\mathfrak{A}$, say via a monomorphism φ, then the dual of φ is a continuous bounded epimorphism from $\mathfrak{U}$ to $\mathfrak{V}$, so that $\mathfrak{V}$ is a continuous bounded homomorphic image of $\mathfrak{U}$. Thus, there is a duality between continuous bounded homomorphic images of a relational space $\mathfrak{U}$ and (isomorphic copies of) subalgebras of the dual algebra $\mathfrak{A}$.

For relational spaces with a single binary relation and Boolean algebras with a single unary operator (modal algebras), a version of the dualities just described is implicit in Goldblatt [12] (see, for example, Theorems 5.9 and 10.9). For more general similarity types, a corresponding version of these dualities follows from Theorems 2.3.1 and 2.3.2 in Goldblatt [13]. (See also Theorem 5.28 in Venema [40].)

Theorems 2.29 and 2.39 strengthen these known duality results by making explicit exactly how the dual structures are constructed. First of all, a duality between ideals and special open sets is constructed— the dual of an ideal M is the special open set H that is the union of the sets in M, and the dual of a special open set H is the ideal M of clopen sets that are included in H—and this duality proves to be a lattice isomorphism (Theorem 2.23). Similarly, a duality between relational congruences and subuniverses is constructed—the dual of a relational congruence Θ is the subuniverse B that consists of clopen sets that are compatible with Θ, and the dual of a subuniverse B is the relational congruence Θ consisting of pairs of elements that belong to the same sets in B—and this duality proves to be a dual lattice isomorphism (Theorem 2.33).

Return now to the relational spaces $\mathfrak{U}$ and $\mathfrak{V}$ with their dual algebras $\mathfrak{A}$ and $\mathfrak{B}$. If $\mathfrak{V}$ is homeo-isomorphic to an inner subspace, say $\mathfrak{W}$, of $\mathfrak{U}$, then the dual algebra $\mathfrak{B}$ is a homomorphic image of $\mathfrak{A}$, and in fact it is isomorphic to the quotient of $\mathfrak{A}$ modulo the ideal that is the dual of the special open set that is the complement of the universe of $\mathfrak{W}$. Conversely, if $\mathfrak{B}$ is a homomorphic image of $\mathfrak{A}$, say via an epimorphism φ, then $\mathfrak{B}$ is isomorphic to the quotient of $\mathfrak{A}$ modulo the ideal M that is the kernel of φ, and therefore the dual structure $\mathfrak{V}$ is homeo-isomorphic to the inner subspace of $\mathfrak{U}$ whose universe is the complement of the special open set that is the dual of the ideal M. This is the content of Theorem 2.29.

Similarly, if $\mathfrak{V}$ is a continuous bounded homomorphic image of $\mathfrak{U}$, say via a continuous bounded epimorphism ϑ, then $\mathfrak{V}$ is homeo-isomorphic to the quotient of $\mathfrak{U}$ modulo the relational congruence Θ that is the kernel of ϑ, and therefore the dual algebra $\mathfrak{B}$ is isomorphic to the subalgebra of $\mathfrak{A}$ whose universe is the dual of the congruence Θ. Conversely, if $\mathfrak{B}$ is isomorphic to a subalgebra, say $\mathfrak{C}$, of $\mathfrak{A}$, then the dual space $\mathfrak{V}$ is homeo-isomorphic to the quotient of $\mathfrak{U}$ modulo the relational congruence that is the dual of the universe of the subalgebra $\mathfrak{C}$, so that $\mathfrak{V}$ is a continuous bounded homomorphic image of $\mathfrak{U}$. This is the content of Theorem 2.39.

2.9 Duality for Completeness

What special properties does the dual space of a complete Boolean algebra with operators possess? (In this section, we do not assume that the operators of a complete algebra are necessarily complete.) The answer is that the space must be *complete* in the topological sense that the closure of every open set is open and hence clopen. The proof of the duality between complete algebras and complete spaces is based on the next lemma, which gives a topological characterization of the suprema that happen to be formable in a not necessarily complete Boolean algebra with operators. The lemma has other applications as well.

Lemma 2.40. *If $(F_i : i \in I)$ is a system of elements (clopen sets) in the dual algebra $\mathfrak{A}$ of a relational space $\mathfrak{U}$, and if $H = \bigcup_i F_i$, then a necessary and sufficient condition for the given system to have a supremum in $\mathfrak{A}$ is that the closure H^- of the set H in $\mathfrak{U}$ be open. If this condition is satisfied, then*

$$\sum_i F_i = H^-,$$

that is to say, the algebraic supremum of the given system is the closure of the set-theoretical union.

Proof. The proof is similar to the proof of the analogous result for Boolean algebras (see Lemma 1 on p. 368 of [10]). Assume first that the supremum F of the given system

$$(F_i : i \in I) \tag{1}$$

exists in $\mathfrak{A}$. The set F belongs to $\mathfrak{A}$, by assumption, so it is a clopen subset of $\mathfrak{U}$, by the definition of $\mathfrak{A}$ as the algebra of clopen subsets of $\mathfrak{U}$. Since F is closed and includes each set F_i, it must include the union H of these sets, and therefore it must also include the closure H^- of this union; in more detail,

$$H \subseteq F \qquad \text{implies} \qquad H^- \subseteq F^- = F.$$

Since F is open, the difference $F - H^-$ is also open. If this difference were non-empty, then it would include a non-empty clopen set G,
because the clopen sets form a base for the topology on $\mathfrak{U}$. The difference $F - G$ would then be a clopen set—and hence an element
in $\mathfrak{A}$—that is properly included in F (because G is non-empty), and
that includes each set F_i (because G is included in $F - H^-$ and is
therefore disjoint from the union H of the sets F_i). Consequently, F
could not be the supremum in $\mathfrak{A}$ of the system of sets in (1), in contradiction to the assumption on F. It follows that $F - H^-$ must be empty,
and consequently $H^- = F$. In other word, H^- is the supremum in $\mathfrak{A}$
of the system in (1), and therefore H^- is open.

If, conversely, the closure H^- is open, then H^- is clearly clopen,
and of course it includes each of the sets F_i, so it is an upper bound
in $\mathfrak{A}$ of the system in (1). If G is any upper bound in $\mathfrak{A}$ of the system
in (1), then G is by definition a clopen set that includes each of the
sets F_i. The union H of the system in (1) must then be included in G,
and therefore the closure of H must be included in G, since G is closed.
Thus, H^- is the least upper bound in $\mathfrak{A}$ of the system in (1). In other
words, the system of sets in (1) has a supremum, and that supremum
is H^-.   □

Theorem 2.41. *The dual algebra of a relational space $\mathfrak{U}$ is (algebraically) complete if and only if $\mathfrak{U}$ is (topologically) complete.*

Proof. Let $\mathfrak{A}$ be the dual algebra of the relational space $\mathfrak{U}$, and assume
first that $\mathfrak{A}$ is complete. An arbitrary open subset H of $\mathfrak{U}$ is the union of
its clopen subsets, because the clopen sets form a base for the topology
of $\mathfrak{U}$. The system of these clopen subsets has a supremum in $\mathfrak{A}$, by the
assumption that $\mathfrak{A}$ is complete. Apply Lemma 2.40 to conclude that
the set H^- must be open. It follows that the space $\mathfrak{U}$ is complete.

Now suppose that the space $\mathfrak{U}$ is complete, and consider an arbitrary
system of elements in $\mathfrak{A}$. The elements in this system are clopen subsets
of $\mathfrak{U}$, because $\mathfrak{A}$—as the dual algebra of $\mathfrak{U}$—consists of the clopen
subsets of $\mathfrak{U}$. Consequently, the union H of this system of elements is
open in $\mathfrak{U}$. The closure H^- must also be open in $\mathfrak{U}$, by the assumption
that $\mathfrak{U}$ is complete. Apply Lemma 2.40 to conclude that the given
system of elements in $\mathfrak{A}$ has a supremum, and that supremum is H^-.
It follows that $\mathfrak{A}$ is complete.   □

A Boolean algebra with operators $\mathfrak{A}$ may possess a certain degree of
completeness without being complete. For instance, $\mathfrak{A}$ is defined to be

countably complete, or *σ-complete*, if the supremum of every countable set of elements in $\mathfrak{A}$ exists. There is a version of Theorem 2.41 that applies to countably complete algebras. To formulate it, we need the notion of a Baire set. A *Baire set* in a Boolean space is a set that belongs to the countably complete Boolean set algebra generated by the set of clopen subsets of the space. In other words, a Baire set is a set that can be obtained from the class of clopen sets by repeated applications of the operations of forming unions and intersections of countable systems of sets. (The operation of forming complements is not really needed because the complement of a countable union—respectively a countable intersection—of sets is the countable intersection—respectively the countable union—of the complements of the sets, by the infinitary versions of the De Morgan laws, and the complement of a clopen set is again a clopen set.) The main topological result about the structure of open Baire sets is that every open Baire set in a Boolean space is the union of countably many clopen sets (see Corollary 1 on p. 375 of [10]). We shall say that a relational space is *countably complete*, or a *σ-space*, if the closure of every open Baire set is open.

Theorem 2.42. *The dual algebra of a relational space $\mathfrak{U}$ is (algebraically) countably complete if and only if $\mathfrak{U}$ is (topologically) countably complete.*

Proof. Let $\mathfrak{A}$ be the dual algebra of the relational space $\mathfrak{U}$, and assume first that $\mathfrak{A}$ is countably complete. If H is an open Baire set in $\mathfrak{U}$, then H must be the union of a countable system of clopen sets, by the remarks preceding the theorem. Since this countable system has a supremum in $\mathfrak{A}$, by the assumed countable completeness of $\mathfrak{A}$, it follows from Lemma 2.40 that H^- is open. Consequently, $\mathfrak{U}$ is a countably complete space, by the definition of such a space.

Assume now that the space $\mathfrak{U}$ is countably complete, and consider any countable system of elements in $\mathfrak{A}$. The elements in this system are clopen subsets of $\mathfrak{U}$ (since they belong to the dual algebra of $\mathfrak{U}$), so their union H is an open set and also a Baire set in $\mathfrak{U}$, by the definition of a Baire set. The closure H^- is therefore open, by the assumed countable completeness of $\mathfrak{U}$. Apply Lemma 2.40 to conclude that the given countable system of elements in $\mathfrak{A}$ has a supremum, and that supremum is H^-. It follows that $\mathfrak{A}$ is a countably complete algebra. $\square$

2.10 Duality for Finite Products

What can one say about the dual space of a direct product of Boolean algebras with operators? This question is easier to answer for products of finite systems than for products of infinite systems: the dual space of the product of finitely many Boolean algebras with operators is homeo-isomorphic to the disjoint union of the dual relational spaces. (A result to this effect is mentioned in passing in Jónsson [19], before Definition 3.2.5, but no details or proofs are given.) Here are the details.

Suppose $(U_i : i \in I)$ is a system of mutually disjoint topological spaces, and U is the union of the sets U_i, for i in I. Every subset F of U can be written in a unique way as a union $F = \bigcup_i F_i$, where F_i is a subset of U_i for each i, namely $F_i = F \cap U_i$; the sets F_i are called the *components* of F. A subset F of U is declared to be open in U if and only if each of its components F_i is open in the corresponding component space U_i. The set of open sets so defined constitutes a topology on U, called the *union topology*, and the resulting topological space is called the *union* of the given system of *component spaces*. Since the complement of a subset F of U is the union of the system of complements of the components F_i (in U_i), it follows that F is closed in U if and only if each component F_i is closed in the component space U_i. Consequently, F is clopen in U if and only if each component is clopen in the corresponding component space.

The union of a disjoint system of topological spaces inherits a number of properties from the component spaces. For instance, the union is a Hausdorff space if and only if each component space is Hausdorff, and the union has a base consisting of clopen sets just in case each component space has a base consisting of clopen sets. If each of the component spaces is compact, then the union is *locally compact* in the sense that every point belongs to the interior of some compact subset. The property of compactness, however, is not inherited by the union unless the given system consists of only finitely many (compact) spaces. If the system is finite in this sense of the word, then the union is a Boolean space if and only if each component space is a Boolean space.

Turn now to the definition of the union of a system of relational spaces.

Definition 2.43. The *union* of a system $(\mathfrak{U}_i : i \in I)$ of mutually disjoint relational spaces is defined to be the structure $\mathfrak{U}$ such that the algebraic part of $\mathfrak{U}$ is the union, in the sense of Definition 1.31, of the relational structures in the given system, and the topology on $\mathfrak{U}$ is the union topology induced by the topologies of the component spaces $\mathfrak{U}_i$. The *disjoint union* of an arbitrary system $(\mathfrak{V}_i : i \in I)$ of relational spaces is defined to be the union of the disjoint system $(\mathfrak{U}_i : i \in I)$ in which $\mathfrak{U}_i$ is the homeo-isomorphic image of the space $\mathfrak{V}_i$ under the function that maps each element r in $\mathfrak{V}_i$ to the element (r, i). $\square$

In connection with the second half of this definition, see the remarks following Definition 1.31.

Lemma 2.44. *If $\mathfrak{U}$ is the union of a disjoint system $(\mathfrak{U}_i : i \in I)$ of relational spaces, then each component space $\mathfrak{U}_i$ is an inner subspace of the union $\mathfrak{U}$ in the sense that the topology on $\mathfrak{U}_i$ is the one inherited from $\mathfrak{U}$, and algebraically $\mathfrak{U}_i$ is an inner substructure of $\mathfrak{U}$. Moreover, a subset of $\mathfrak{U}_i$ is open, closed, clopen, or compact in $\mathfrak{U}$ if and only if it has the same property in $\mathfrak{U}_i$.*

Proof. Focus on the case of a ternary relation R in $\mathfrak{U}$. This relation is, by definition, the disjoint union of the corresponding relations, say R_i, in $\mathfrak{U}_i$, for i in I. In particular, the restriction of R to the universe of $\mathfrak{U}_i$ must coincide with the relation R_i. For the same reason, if t is an element in $\mathfrak{U}_i$, and if $R(r, s, t)$ holds in $\mathfrak{U}$, then the triple (r, s, t) must belong to the relation R_i, so that r and s must be in $\mathfrak{U}_i$. Thus, $\mathfrak{U}_i$ is algebraically an inner substructure of $\mathfrak{U}$, by the definition of such a substructure.

If F is any open subset of $\mathfrak{U}$, then for each index i, the component of F in $\mathfrak{U}_i$, which is just the intersection $F \cap U_i$, is open in $\mathfrak{U}_i$, by the definition of the union topology on $\mathfrak{U}$. Conversely, if G is any open subset of $\mathfrak{U}_i$, and if for each index j, the set F_j is defined by

$$F_j = \begin{cases} G & \text{if } i = j, \\ \varnothing & \text{if } i \neq j, \end{cases}$$

then the union $F = \bigcup_j F_j$ is open in $\mathfrak{U}$, since each component of this union is open in the relevant component space. Obviously,

$$G = F = F \cap U_i,$$

so every open set in $\mathfrak{U}_i$ is the intersection with U_i of an open set in $\mathfrak{U}$. Conclusion: the topology on $\mathfrak{U}_i$ is the one inherited from $\mathfrak{U}$.

The preceding argument also shows that every open set G in $\mathfrak{U}_i$ is actually equal to an open set F in $\mathfrak{U}$. Conversely, if F is any open set in $\mathfrak{U}$ that is a subset of $\mathfrak{U}_i$, then F must coincide with its component in $\mathfrak{U}_i$ and therefore must be open in $\mathfrak{U}_i$, by the definition of the union topology. Thus, a subset of $\mathfrak{U}_i$ is open in $\mathfrak{U}_i$ if and only if it is open in $\mathfrak{U}$. Similar arguments apply to closed sets, clopen sets, and compact sets. $\square$

The disjoint union of a system of relational spaces is almost a relational space. The compactness property may fail, but the union is at any rate locally compact.

Lemma 2.45. *Suppose $\mathfrak{U}$ is the union of a disjoint system of relational spaces. The union topology turns the universe of $\mathfrak{U}$ into a locally compact Hausdorff space in which the clopen sets form a base for the topology. Under this topology, the relations in $\mathfrak{U}$ are clopen and continuous.*

Proof. Let $\mathfrak{U}$ be the union of a disjoint system $(\mathfrak{U}_i : i \in I)$ of relational spaces. A subset of $\mathfrak{U}_i$ is open, closed, clopen, or compact in $\mathfrak{U}$ if and only if it is open, closed, clopen, or compact in $\mathfrak{U}_i$, by Lemma 2.44. One consequence of this observation is that the clopen sets in $\mathfrak{U}$ form a base for the topology on $\mathfrak{U}$. In fact, the clopen subsets of the component spaces form a base for the topology on $\mathfrak{U}$. A second consequence is that $\mathfrak{U}$ is a Hausdorff space. Indeed, two points belonging to the same component space $\mathfrak{U}_i$ are separated by a clopen subset of $\mathfrak{U}_i$, while two points belonging to distinct component spaces $\mathfrak{U}_i$ and $\mathfrak{U}_j$ are separated by the clopen sets U_i and U_j; and all these separating sets remain clopen in $\mathfrak{U}$. A third consequence is that $\mathfrak{U}$ is locally compact. In fact, if r is any point in $\mathfrak{U}$, then r belongs to one of the component spaces $\mathfrak{U}_i$, and U_i is an open compact set (both in $\mathfrak{U}_i$ and in $\mathfrak{U}$) that contains r. Thus, $\mathfrak{U}$ has the topology of a locally compact Hausdorff space in which the clopen sets form a base for the topology.

The next task is to show that the relations in $\mathfrak{U}$ are clopen and those of rank at least two are continuous. Focus on the case of a ternary relation R that is the disjoint union of the corresponding ternary relations R_i in $\mathfrak{U}_i$, for i in I. To prove that R is clopen, it must be shown that for any two clopen sets F and G in $\mathfrak{U}$, the image set

$$R^*(F \times G) = \{t \in U : R(r, s, t) \text{ for some } r \in F \text{ and } s \in G\} \tag{1}$$

is clopen. Write F and G as the unions of their components,

$$F = \bigcup_i F_i \quad \text{and} \quad G = \bigcup_i G_i. \tag{2}$$

A set in $\mathfrak{U}$ is clopen if and only if each component of the set is clopen in the relevant component space, so the components F_i and G_i must be clopen sets in $\mathfrak{U}_i$ for each i. Consequently, the set

$$R_i^*(F_i \times G_i) = \{t \in U_i : R_i(r, s, t) \text{ for some } r \in F_i \text{ and } s \in G_i\} \tag{3}$$

is clopen in $\mathfrak{U}_i$ for each i, by the assumption that $\mathfrak{U}_i$ is a relational space. In order to prove that (1) is clopen, it therefore suffices to prove that

$$R^*(F \times G) = \bigcup_i R_i^*(F_i \times G_i), \tag{4}$$

by the definition of the union topology.

The equality in (4) follows rather easily from the fact that

$$R = \bigcup_i R_i \tag{5}$$

is a disjoint union. Indeed, consider an element t in $\mathfrak{U}$. If t belongs to the left side of (4), then there must be elements r in F and s in G such that $R(r, s, t)$ holds (in $\mathfrak{U}$), by (1). In view of (5), we must have $R_i(r, s, t)$ (in $\mathfrak{U}_i$) for some index i, so the elements r and s are in $\mathfrak{U}_i$, and therefore they belong to the sets

$$F \cap U_i = F_i \quad \text{and} \quad G \cap U_i = G_i$$

respectively. In view of (3), it follows that t belongs to $R_i^*(F_i \times G_i)$, so t belongs to the right side of (4). Thus, the left side of (4) is included in the right side.

To establish the reverse inclusion, assume that t belongs to the right side of (4). In this case, t is in $R_i^*(F_i \times G_i)$ for some index i. Consequently, there must be elements r in F_i and s in G_i such that $R_i(r, s, t)$ holds. Clearly, r is in F and s in G, by (2), and $R(r, s, t)$ holds in $\mathfrak{U}$, by (5), so t must belong to the left side of (4), by (1). Thus, the right side of (4) is included in the left. This establishes (4) and proves that the relation R is clopen.

To prove that R is continuous, consider an open subset H of $\mathfrak{U}$. It must be shown that the set

$$R^{-1}(H) = \{(r, s) \in U \times U : R(r, s, t) \text{ implies } t \in H\} \tag{6}$$

is open in the product topology on $U \times U$. Write H as the union of its components,

$$H = \bigcup_i H_i. \tag{7}$$

A set is open in $\mathfrak{U}$ if and only if each of its components is open in the relevant component space, so for each index i, the component H_i is open in $\mathfrak{U}_i$. Consequently, the set

$$R_i^{-1}(H_i) = \{(r, s) \in U_i \times U_i : R_i(r, s, t) \text{ implies } t \in H_i\} \tag{8}$$

is open in the product topology on $U_i \times U_i$—and therefore open in the product topology on $U \times U$—by the assumption that $\mathfrak{U}_i$ is a relational space. Also, each of the sets $U_i \times U_j$ is open in $U \times U$, since U_i and U_j are open subsets of $\mathfrak{U}$. Consequently, the union

$$G = \left(\bigcup_i R_i^{-1}(H_i)\right) \cup \left(\bigcup\{U_i \times U_j : i, j \in I \text{ and } i \neq j\}\right) \tag{9}$$

is open in $U \times U$.

In view of the preceding observations, it suffices to prove that

$$R^{-1}(H) = G. \tag{10}$$

Consider a pair (r, s) in $U \times U$. There are two possibilities: either r and s belong to the same component space or they belong to distinct component spaces. If they belong to distinct component spaces, then the pair (r, s) obviously belongs to G, by (9); and the pair vacuously belongs to $R^{-1}(H)$, by (6), because the relation $R(r, s, t)$ can never hold in $\mathfrak{U}$, by (5). Thus, $R^{-1}(H)$ and G contain the same pairs having coordinates in distinct component spaces.

Suppose now that r and s belong to the same component space, say $\mathfrak{U}_i$, with the goal of showing that the pair (r, s) belongs to the set $R^{-1}(H)$ if and only if it belongs to G. Assume first that (r, s) belongs to $R^{-1}(H)$. In order to prove that (r, s) belongs to $R_i^{-1}(H_i)$, and therefore to G, it must be shown that the hypothesis $R_i(r, s, t)$ implies that t belongs to H_i, by (8). The hypothesis implies that $R(r, s, t)$ must hold, by (5), and therefore t must belong to H, by (6). The hypothesis also implies that t must be in U_i, so t belongs to the intersection $H \cap U_i$, which is just H_i. Assume now that (r, s) belongs to G. Since r and s are assumed to be in the same component space $\mathfrak{U}_i$, the pair (r, s) must belong to $R_i^{-1}(H_i)$, by (9). In order to prove that (r, s) belongs to $R^{-1}(H)$, it must be shown that the hypothesis $R(r, s, t)$ implies t is in H. The hypothesis and the assumption that r and s are in $\mathfrak{U}_i$

imply that $R_i(r, s, t)$ holds, by (5); so t belongs to H_i, by (8), and therefore t belongs to H, by (7). This completes the proof of (10) and shows that R is continuous. □

The lemma implies that the disjoint union of a system of relational spaces possesses all of the properties of a relational space except perhaps compactness. Since the union of finitely many compact spaces is compact, we arrive at the following corollary.

Corollary 2.46. *The disjoint union of finitely many relational spaces is again a relational space.*

The next theorem describes the duality that exists between unions of finitely many disjoint relational spaces and the product of their dual algebras.

Theorem 2.47. *The dual algebra of the union of a finite system of disjoint relational spaces is equal to the internal product of the dual algebras of the system.*

Proof. Let $\mathfrak{U}$ be the union of a finite system

$$(\mathfrak{U}_i : i \in I) \tag{1}$$

of mutually disjoint relational spaces, and let $\mathfrak{A}$ and $\mathfrak{A}_i$ be the dual algebras of $\mathfrak{U}$ and $\mathfrak{U}_i$ respectively. It is to be shown that $\mathfrak{A}$ is the internal product of the system

$$(\mathfrak{A}_i : i \in I) \tag{2}$$

in the sense defined in Section 1.10 (see also the remarks following Definition 2.48 in the next section). The complex algebra $\mathfrak{Cm}(U)$ is equal to the internal product of the system of complex algebras

$$(\mathfrak{Cm}(U_i) : i \in I), \tag{3}$$

by Theorem 1.32. For each index i, the algebra $\mathfrak{A}_i$ is a subalgebra of $\mathfrak{Cm}(U_i)$, by its very construction, so the internal product of the system in (2) is a subalgebra of the internal product of the system in (3). Consequently, the internal product of the system in (2) is a subalgebra of $\mathfrak{Cm}(U)$. The universe of this internal product consists of the sets of the form

$$F = \bigcup_i F_i, \tag{4}$$

where for each i, the set F_i belongs to $\mathfrak{A}_i$, that is to say, F_i is a clopen subset of $\mathfrak{U}_i$.

The algebra $\mathfrak{A}$ is also a subalgebra of $\mathfrak{Cm}(U)$, and its universe is the set of clopen subsets of $\mathfrak{U}$. Since $\mathfrak{U}$ is the union of the spaces in (1), the clopen subsets of $\mathfrak{U}$ are just the sets of the form (4), where F_i is a clopen subset of $\mathfrak{U}_i$ for each i. Thus, the universe of $\mathfrak{A}$ coincides with the universe of the internal product of the system in (2). Since $\mathfrak{A}$ and the internal product are both subalgebras of $\mathfrak{Cm}(U)$, and they have the same universe, they must be the same subalgebra. □

The dual version of the preceding theorem says that the dual space of the direct product of a finite system of Boolean algebras with operators is homeo-isomorphic to the disjoint union of the system of dual relational spaces.

2.11 Duality for Subdirect Products

The description of the dual space of a direct product of infinitely many Boolean algebras with operators is more involved. There is in fact a dual correspondence between certain subalgebras of such products and the compactifications of unions of relational spaces. In view of Lemma 2.45, the disjoint union of an infinite system of relational spaces possesses all the properties of a relational space except possibly compactness, which has been replaced by a weaker property, namely local compactness. Let us call such a structure a *locally compact* relational space. Thus, a locally compact relational space is not required to be compact, whereas a relational space is compact by its very definition.

Definition 2.48. A *compactification* of a locally compact relational space $\mathfrak{U}$ is defined to be a relational space $\mathfrak{V}$ such that $\mathfrak{U}$ is a dense inner subspace of $\mathfrak{V}$ in the following sense: algebraically, $\mathfrak{U}$ is an inner substructure of $\mathfrak{V}$; the topology on $\mathfrak{U}$ is the one inherited from $\mathfrak{V}$; and the (topological) closure of the set U in $\mathfrak{V}$ is just the set V. □

Our immediate goal is a description of the relationship between compactifications of disjoint unions of infinite systems relational spaces, and certain subalgebras of direct products of infinite systems of Boolean algebras with operators. We proceed to establish the notation that will be used in this description. Fix a disjoint system $(\mathfrak{U}_i : i \in I)$ of relational spaces for the remainder of this section, and let $\mathfrak{U}$ the union of this system. For each index i, let $\mathfrak{A}_i$ be the dual algebra of the space $\mathfrak{U}_i$, and let $\mathfrak{A}$ be the internal product of the system of dual algebras. The elements in $\mathfrak{A}$ are the subsets of $\mathfrak{U}$ of the form $F = \bigcup_i F_i$, where each set F_i is an element in $\mathfrak{A}_i$, that is to say, F_i is a clopen subset of $\mathfrak{U}_i$. The operations of $\mathfrak{A}$ are performed coordinatewise: if

$$F = \bigcup_i F_i \qquad \text{and} \qquad G = \bigcup_i G_i$$

are elements in $\mathfrak{A}$, then

$$F + G = \bigcup_i (F_i + G_i) \qquad \text{and} \qquad -F = \bigcup_i -F_i,$$

and if f is an operator of rank n, and if

$$F_0 = \bigcup_i F_{0,i} , \qquad \ldots \qquad , F_{n-1} = \bigcup_i F_{n-1,i}$$

is a sequence of elements in $\mathfrak{A}$, then

$$f(F_0, \ldots, F_{n-1}) = \bigcup_i f(F_{0,i}, \ldots, F_{n-1,i}),$$

where the operations on the left sides of the equations above are those of $\mathfrak{A}$, while the ones on the right sides, after the union symbols, are the operations of the factor algebras $\mathfrak{A}_i$.

Lemma 2.49. *If $\mathfrak{V}$ is a compactification of $\mathfrak{U}$, then each component space $\mathfrak{U}_i$ is an inner subspace of $\mathfrak{V}$ in the sense of Definition 2.26. The universe of $\mathfrak{U}$ is an open subset of $\mathfrak{V}$. A subset of $\mathfrak{U}_i$ is open, closed, clopen, or compact in $\mathfrak{V}$ if and only if it has this same property in $\mathfrak{U}_i$.*

Proof. For each index i, the relational space $\mathfrak{U}_i$ is algebraically an inner substructure of $\mathfrak{U}$, by Lemma 2.44, and $\mathfrak{U}$ is an inner substructure of $\mathfrak{V}$, by Definition 2.48, so $\mathfrak{U}_i$ is an inner substructure of $\mathfrak{V}$, by the transitivity of the relation of being an inner substructure. Similarly, the universe of $\mathfrak{U}_i$ is topologically a subspace of the universe of $\mathfrak{U}$, by Lemma 2.44, and the universe of $\mathfrak{U}$ is a topologically a subspace of the universe of $\mathfrak{V}$, by Definition 2.48, so the universe of $\mathfrak{U}_i$ is topologically a subspace of the universe of $\mathfrak{V}$, by the transitivity of the relation of being a topological subspace. Combine these observations to conclude that $\mathfrak{U}_i$ is a subspace of $\mathfrak{V}$ in the sense of Definition 2.26.

Every dense, locally compact subspace of a Hausdorff space is open in that space (see Corollary 1 on p. 400 of [10]). Since the universe of $\mathfrak{U}$ is topologically a dense, locally compact subspace of $\mathfrak{V}$, it follows that this universe must be open in the topology of $\mathfrak{V}$.

The universe of $\mathfrak{U}_i$ is open in $\mathfrak{U}$, by Lemma 2.44, and the universe of $\mathfrak{U}$ is open in $\mathfrak{V}$, by the observations of the preceding paragraph, so the universe of $\mathfrak{U}_i$ is also open in $\mathfrak{V}$. Consequently, a subset of $\mathfrak{U}_i$ is open in $\mathfrak{V}$ if and only if it is open in $\mathfrak{U}_i$. From this it also follows that a subset F of $\mathfrak{U}_i$ is compact in $\mathfrak{V}$ if and only if it is compact in $\mathfrak{U}_i$. The reason is that every open cover of F in $\mathfrak{U}_i$ remains an open cover of F in $\mathfrak{V}$, and inversely, the intersection with the universe of $\mathfrak{U}_i$ of any open cover of F in $\mathfrak{V}$ yields an open cover of F in $\mathfrak{U}_i$. In compact Hausdorff spaces, the closed sets coincide with the compact sets. Consequently, a subset of $\mathfrak{U}_i$ is closed in $\mathfrak{V}$ if and only if it is closed in $\mathfrak{U}_i$, by the preceding remark. Combine these observations to conclude that a subset of $\mathfrak{U}_i$ is clopen in $\mathfrak{V}$ if and only if it is clopen in $\mathfrak{U}_i$. $\quad\square$

There is a subalgebra of the internal product $\mathfrak{A}$ that will play a special role in our discussion, namely the one that is generated by the union $\bigcup_i A_i$ of the universes of the factor algebras. The elements of this subalgebra are precisely those sets $F = \bigcup_i F_i$ in $\mathfrak{A}$ such that the system $(F_i : i \in I)$ is *constant almost everywhere* in the following sense: there is a term τ in the language of Boolean algebras with operators that is built up from the distinguished constant symbols (symbols for operations of rank 0, including 0 and 1) and the operation symbols, without using any variables, such that F_i coincides with the value of τ in $\mathfrak{A}_i$ for all but finitely many indices i. We denote this subalgebra by $\mathfrak{D}$ and call it the *weak internal product* of the system $(\mathfrak{A}_i : i \in I)$. We now prove that the dual of every compactification of $\mathfrak{U}$ corresponds to a subalgebra of $\mathfrak{A}$ that includes $\mathfrak{D}$.

Lemma 2.50. *If $\mathfrak{B}$ is the dual algebra of a compactification of $\mathfrak{U}$, then the set*

$$B_0 = \{F \cap U : F \in B\}$$

is a subuniverse of $\mathfrak{A}$ that includes the universe of $\mathfrak{D}$. Moreover, $\mathfrak{B}$ is isomorphic to the corresponding subalgebra $\mathfrak{B}_0$ via the function that maps F to $F \cap U$ for each F in $\mathfrak{B}$.

Proof. Let $\mathfrak{V}$ be a compactification of $\mathfrak{U}$, and $\mathfrak{B}$ the dual algebra of $\mathfrak{V}$. Algebraically, $\mathfrak{U}$ is an inner substructure of $\mathfrak{V}$, by Definition 2.48, so the identity function ϑ on $\mathfrak{U}$ is a bounded monomorphism from $\mathfrak{U}$ to $\mathfrak{V}$, by Corollary 1.14. The algebraic dual of ϑ is, by Theorem 1.9, the complete epimorphism ψ from $\mathfrak{Cm}(V)$ to $\mathfrak{Cm}(U)$ that is defined by

$$\psi(F) = \vartheta^{-1}(F)$$

for elements F in $\mathfrak{Cm}(V)$, that is to say, for subsets F of $\mathfrak{V}$. Since

$$\vartheta^{-1}(F) = \{r \in U : \vartheta(r) \in F\} = \{r \in U : r \in F\} = F \cap U,$$

by the definition of the inverse image under ϑ of a set, and the definition of ϑ as the identity function on $\mathfrak{U}$, it may be concluded that

$$\psi(F) = F \cap U \qquad\qquad (1)$$

for each F in $\mathfrak{Cm}(V)$. The dual algebra $\mathfrak{B}$ is, by construction, a subalgebra of $\mathfrak{Cm}(V)$. The appropriate restriction of ψ therefore maps $\mathfrak{B}$ onto a subalgebra of $\mathfrak{Cm}(U)$, namely the subalgebra $\mathfrak{B}_0$ with universe B_0, by (1) and the homomorphism properties of ψ. We shall also use the symbol ψ to refer to this restriction.

The dual algebra $\mathfrak{A}_i$ of the relational space $\mathfrak{U}_i$ is a subalgebra of $\mathfrak{Cm}(U_i)$, by construction. The internal product $\mathfrak{A}$ of the system

$$(\mathfrak{A}_i : i \in I) \qquad\qquad (2)$$

is therefore a subalgebra of the internal product of the system

$$(\mathfrak{Cm}(U_i) : i \in I).$$

Since the internal product of the latter system is just $\mathfrak{Cm}(U)$, by Theorem 1.32, it follows that $\mathfrak{A}$ is a subalgebra of $\mathfrak{Cm}(U)$. We proceed to show that B_0 is a subset of the universe of $\mathfrak{A}$. Since $\mathfrak{B}_0$ and $\mathfrak{A}$ are both subalgebras of $\mathfrak{Cm}(U)$, it then follows that $\mathfrak{B}_0$ is a subalgebra of $\mathfrak{A}$.

An arbitrary element in B_0 has the form $F \cap U$ for some set F in $\mathfrak{B}$, by the definition of B_0. The set F is clopen in $\mathfrak{V}$, by the definition of $\mathfrak{B}$ as the dual of $\mathfrak{V}$, and the universe U_i of $\mathfrak{U}_i$ is clopen in $\mathfrak{V}$, by Lemma 2.49, so the intersection of F with U_i, the set

$$F_i = F \cap U_i,$$

is clopen in $\mathfrak{V}$ and therefore also in $\mathfrak{U}_i$, by Lemma 2.49. Consequently, F_i belongs to $\mathfrak{A}_i$, by the definition of $\mathfrak{A}_i$ as the dual algebra of $\mathfrak{U}_i$. Since

$$F \cap U = F \cap (\textstyle\bigcup_i U_i) = \textstyle\bigcup_i (F \cap U_i) = \textstyle\bigcup_i F_i,$$

it follows from the definition of $\mathfrak{A}$ as the internal product of the system in (2) that $F \cap U$ belongs to $\mathfrak{A}$. Thus, B_0 is a subset of $\mathfrak{A}$, so $\mathfrak{B}_0$ is a subalgebra of $\mathfrak{A}$.

The next step is to show that $\mathfrak{D}$ is a subalgebra of $\mathfrak{B}_0$. Since both of these algebras are subalgebras of $\mathfrak{A}$, it suffices to show that every element in a set of generators of $\mathfrak{D}$ belongs to $\mathfrak{B}_0$. The algebra $\mathfrak{D}$ is, by definition, generated by the union of the universes of the algebras $\mathfrak{A}_i$, and the elements in these universes are just the clopen subsets of the component spaces $\mathfrak{U}_i$. An arbitrary clopen subset F of $\mathfrak{U}_i$ remains clopen in $\mathfrak{V}$, by Lemma 2.49, and therefore belongs to $\mathfrak{B}$. Since the intersection of F with U is just F, it follows that F belongs to $\mathfrak{B}_0$. Thus, every element in a set of generators for $\mathfrak{D}$ does belong to $\mathfrak{B}_0$, so $\mathfrak{D}$ is a subalgebra of $\mathfrak{B}_0$.

The function ψ maps $\mathfrak{B}$ homomorphically onto $\mathfrak{B}_0$. In order to show that the two algebras are isomorphic, it suffices to prove that ψ restricted to $\mathfrak{B}$ is one-to-one. Suppose F and G are elements in $\mathfrak{B}$. Observe that

$$(F \cap U)^- = F \quad \text{and} \quad (G \cap U)^- = G, \tag{3}$$

where X^- denotes the topological closure in $\mathfrak{V}$ of a subset X of $\mathfrak{V}$. In more detail, $(F \cap U)^- = F \cap U^-$ because F is a clopen set (see Exercise 13(e) on p. 62 of [10]), and therefore

$$(F \cap U)^- = F \cap U^- = F \cap V = F,$$

because U is dense in $\mathfrak{V}$, and F is a subset of $\mathfrak{V}$. If $\psi(F) = \psi(G)$, then

$$F \cap U = G \cap U,$$

by (1), and therefore

$$F = (F \cap U)^- = (G \cap U)^- = G,$$

by (3). Thus, ψ is one-to-one. $\square$

We shall refer to the algebra $\mathfrak{B}_0$ in Lemma 2.50 as the *relativization of $\mathfrak{B}$ to U*, and we shall call the isomorphism ψ from $\mathfrak{B}$ to $\mathfrak{B}_0$ the *relativization isomorphism*.

The next task is to determine how the subalgebras of $\mathfrak{A}$ that correspond to various compactifications of $\mathfrak{U}$ are related to one another.

Lemma 2.51. *Suppose $\mathfrak{V}$ and $\mathfrak{W}$ are compactifications of $\mathfrak{U}$, with dual algebras $\mathfrak{B}$ and $\mathfrak{C}$ respectively. The relativization of $\mathfrak{C}$ to U is a subalgebra of the relativization of $\mathfrak{B}$ to U if and only if there is a continuous bounded epimorphism from $\mathfrak{V}$ to $\mathfrak{W}$ that is the identity function on U.*

Proof. Let $\mathfrak{B}_0$ be the relativization of $\mathfrak{B}$ to U, and ψ the corresponding relativization isomorphism defined by

$$\psi(F) = F \cap U \tag{1}$$

for each set F in $\mathfrak{B}$. Similarly, let $\mathfrak{C}_0$ be the relativization of $\mathfrak{C}$ to U, and ϱ the corresponding relativization isomorphism defined by

$$\varrho(F) = F \cap U \tag{2}$$

for each set F in $\mathfrak{C}$.

Assume first that there is a continuous bounded epimorphism ϑ from $\mathfrak{V}$ to $\mathfrak{W}$ that maps each element in $\mathfrak{U}$ to itself. The dual of ϑ is the monomorphism φ from $\mathfrak{C}$ to $\mathfrak{B}$ that is defined by

$$\varphi(F) = \vartheta^{-1}(F) = \{r \in V : \vartheta(r) \in F\} \tag{3}$$

for every set F in $\mathfrak{C}$, by Theorem 2.16. The composition

$$\delta = \psi \circ \varphi \circ \varrho^{-1} \tag{4}$$

is a monomorphism from $\mathfrak{C}_0$ to $\mathfrak{B}_0$ (see the diagram below).

$$
\begin{array}{ccc}
\mathfrak{B} \xleftarrow{\ \varphi\ } \mathfrak{C} & \quad & \mathfrak{V} \xrightarrow{\ \vartheta\ } \mathfrak{W} \\
\psi \downarrow \qquad \downarrow \varrho & & \\
\mathfrak{B}_0 \xleftarrow{\ \delta\ } \mathfrak{C}_0 & &
\end{array}
$$

We proceed to show that δ is the identity function on $\mathfrak{C}_0$. From this it follows at once that $\mathfrak{C}_0$ is a subalgebra of $\mathfrak{B}_0$. For each element G in $\mathfrak{C}_0$, there is a unique element F in $\mathfrak{C}$ such that

$$G = \varrho(F) = F \cap U, \tag{5}$$

by (2) and the fact that ϱ is an isomorphism from $\mathfrak{C}$ to $\mathfrak{C}_0$. An easy computation yields

$$\delta(G) = \psi(\varphi(\varrho^{-1}(G))) = \psi(\varphi(F))$$
$$= \psi(\vartheta^{-1}(F)) = \vartheta^{-1}(F) \cap U = F \cap U = G.$$

The first equality follows from (4), the second and final equalities from (5), the third equality from (3), and the fourth equality from (1).

For the fifth equality, observe that an element r is in $\vartheta^{-1}(F) \cap U$ just in case r is in U and $\vartheta(r)$ is in F, by (3). Since ϑ is the identity function on U, this last condition is equivalent to saying that r is in $F \cap U$.

To prove the converse direction of the lemma, assume that $\mathfrak{C}_0$ is a subalgebra of $\mathfrak{B}_0$. The identity function on $\mathfrak{C}_0$—call it δ—is then a monomorphism from $\mathfrak{C}_0$ to $\mathfrak{B}_0$. The composition

$$\varphi = \psi^{-1} \circ \delta \circ \varrho \tag{6}$$

is a monomorphism from $\mathfrak{C}$ to $\mathfrak{B}$ (see the diagram above). For each set F belonging to any one of the factor algebras $\mathfrak{A}_i$, we have

$$\varphi(F) = \psi^{-1}(\delta(\varrho(F))) = \psi^{-1}(\delta(F \cap U)) = \psi^{-1}(F \cap U) = F,$$

by (6), (2), the assumption that δ is the identity function on $\mathfrak{C}_0$, and (1). Thus, φ is the identity function on the universe of $\mathfrak{A}_i$ for each index i.

The dual of φ is the continuous bounded epimorphism ϑ from $\mathfrak{V}$ to $\mathfrak{W}$ that is determined by

$$\vartheta(r) \in F \qquad \text{if and only if} \qquad r \in \varphi(F) \tag{7}$$

for all elements r in $\mathfrak{V}$ and sets F in $\mathfrak{C}$, by Theorem 2.20. It remains to show that ϑ is the identity function on $\mathfrak{U}$. An element r in $\mathfrak{U}$ necessarily belongs to one of the spaces $\mathfrak{U}_i$, by the definition of $\mathfrak{U}$ as the union of these spaces, and φ is the identity function on the sets in $\mathfrak{A}_i$, by the observations of the preceding paragraph. Consequently, for such an element r, and for sets F in $\mathfrak{A}_i$, the equivalence in (7) assumes the form

$$\vartheta(r) \in F \qquad \text{if and only if} \qquad r \in F. \tag{8}$$

The set X of all sets in $\mathfrak{A}_i$ that contain the element r is an ultrafilter in $\mathfrak{A}_i$, and r is the only element in $\mathfrak{U}_i$ that belongs to each of the sets in X (since two distinct elements in $\mathfrak{U}_i$ are always separated by a clopen set). On the other hand, the sets in X all contain $\vartheta(r)$, by (8), so we must have $\vartheta(r) = r$. $\quad\square$

Notice that the argument in the final paragraph proves a somewhat stronger assertion: if the dual of a continuous bounded homomorphism ϑ from $\mathfrak{V}$ to $\mathfrak{W}$ is the identity function on elements in $\mathfrak{A}_i$ for each index i, then ϑ is the identity function on elements in $\mathfrak{U}$.

Lemma 2.52. *Suppose $\mathfrak{V}$ and $\mathfrak{W}$ are compactifications of $\mathfrak{U}$, with dual algebras $\mathfrak{B}$ and $\mathfrak{C}$ respectively. The spaces $\mathfrak{V}$ and $\mathfrak{W}$ are homeo-isomorphic via a function that is the identity function on $\mathfrak{U}$ if and only if the relativizations of $\mathfrak{B}$ and $\mathfrak{C}$ to U are equal, or equivalently, if and only if $\mathfrak{B}$ and $\mathfrak{C}$ are isomorphic via a function that is the identity function on $\mathfrak{A}_i$ for each i.*

Proof. Write $\mathfrak{B}_0$ and $\mathfrak{C}_0$ for the respective relativizations of $\mathfrak{B}$ and $\mathfrak{C}$ to the set U. If there is a homeo-isomorphism from $\mathfrak{V}$ to $\mathfrak{W}$ that is the identity function on $\mathfrak{U}$, then the relativizations $\mathfrak{B}_0$ and $\mathfrak{C}_0$ are subalgebras of each other, by Lemma 2.51, and are therefore equal.

Suppose now that $\mathfrak{B}_0$ and $\mathfrak{C}_0$ are equal. There are then continuous bounded epimorphisms ϑ from $\mathfrak{V}$ to $\mathfrak{W}$, and δ from $\mathfrak{W}$ to $\mathfrak{V}$, that are the identity function on $\mathfrak{U}$, by Lemma 2.51. The composition $\delta \circ \vartheta$ is a continuous bounded epimorphism from $\mathfrak{V}$ to $\mathfrak{V}$ that is the identity function on $\mathfrak{U}$, and the identity function on $\mathfrak{V}$ is also a continuous bounded epimorphism from $\mathfrak{V}$ to $\mathfrak{V}$ that is the identity function on $\mathfrak{U}$. The universe of $\mathfrak{U}$ is a dense subset of $\mathfrak{V}$, by Definition 2.48. Two continuous functions from a topological space to a Hausdorff space that agree on a dense subset agree on the whole space (see Corollary 2 on p. 315 of [10]), so $\delta \circ \vartheta$ must be the identity function on $\mathfrak{V}$. A symmetric argument shows that $\vartheta \circ \delta$ is the identity function on $\mathfrak{W}$. It follows that ϑ is a bijection and δ its inverse.

Since each of ϑ and its inverse δ is a continuous bounded epimorphism, the function ϑ must be a homeo-isomorphism from $\mathfrak{V}$ to $\mathfrak{W}$. For example, consider the case of a ternary relation R. If $R(r, s, t)$ holds in $\mathfrak{V}$, then $R(\vartheta(r), \vartheta(s), \vartheta(t))$ must hold in $\mathfrak{W}$, by the homomorphism properties of ϑ. Conversely, if $R(\vartheta(r), \vartheta(s), \vartheta(t))$ holds in $\mathfrak{W}$, then $R(\delta(\vartheta(r)), \delta(\vartheta(s)), \delta(\vartheta(t)))$ must hold in $\mathfrak{V}$, by the homomorphism properties of δ. Since $\delta \circ \vartheta$ is the identity function on $\mathfrak{V}$, it may be concluded that $R(r, s, t)$ holds in $\mathfrak{V}$. Consequently, the function ϑ isomorphically preserves the relation R.

To establish the second equivalence of the lemma, assume that $\mathfrak{C}$ is isomorphic to $\mathfrak{B}$ via a function φ that is the identity function on each algebra $\mathfrak{A}_i$. The dual of φ is a homeo-isomorphism ϑ from $\mathfrak{V}$ to $\mathfrak{W}$, by Corollary 2.19, and ϑ is the identity function on $\mathfrak{U}$, by the remark following Lemma 2.51.

Assume now that $\mathfrak{V}$ is homeo-isomorphic to $\mathfrak{W}$ via a function ϑ that is the identity function on $\mathfrak{U}$. The relativizations $\mathfrak{B}_0$ and $\mathfrak{C}_0$ are then equal, by the first part of the lemma. If ψ is the relativization isomorphism from $\mathfrak{B}$ to $\mathfrak{B}_0$, and ϱ the relativization isomorphism from $\mathfrak{C}$ to $\mathfrak{C}_0$, then the composition

$$\varphi = \psi^{-1} \circ \varrho$$

is an isomorphism from $\mathfrak{C}$ to $\mathfrak{B}$, and it is easy to check that φ is the identity function on each of the algebras $\mathfrak{A}_i$. Indeed, if F is a set in $\mathfrak{A}_i$, then

$$\varphi(F) = \psi^{-1}(\varrho(F)) = \psi^{-1}(F \cap U) = F,$$

by the definitions of ψ and ϱ. $\square$

The final lemma says that every algebra between $\mathfrak{D}$ and $\mathfrak{A}$ comes from the dual of some compactification of $\mathfrak{U}$.

Lemma 2.53. *Every subalgebra of $\mathfrak{A}$ that includes $\mathfrak{D}$ is the relativization to U of the dual algebra of some compactification of $\mathfrak{U}$.*

Proof. Consider a subalgebra $\mathfrak{C}$ of $\mathfrak{A}$ that includes $\mathfrak{D}$, and let $\mathfrak{V}$ be the dual space of $\mathfrak{C}$. The idea of the proof is to construct a dense inner subspace of $\mathfrak{V}$ (in the sense of Definition 2.48) that is the image of $\mathfrak{U}$ under a homeo-isomorphism ϱ, and that has the property that the relativization of the dual algebra of $\mathfrak{V}$ to this dense inner subspace is the image of $\mathfrak{C}$ under the isomorphism induced by ϱ. An application of the general algebraic Exchange Principle then yields a compactification of $\mathfrak{U}$ with the property that the dual algebra of the compactification, when relativized to U, coincides with the given subalgebra $\mathfrak{C}$.

Because $\mathfrak{V}$ is assumed to be the dual space of $\mathfrak{C}$, the topology on $\mathfrak{V}$ is the one induced by $\mathfrak{C}$. Thus, the elements in $\mathfrak{V}$ are the ultrafilters of elements in $\mathfrak{C}$, the clopen subsets of $\mathfrak{V}$ are the sets of the form

$$G_F = \{Y \in V : F \in Y\}, \tag{1}$$

where F ranges over the elements in $\mathfrak{C}$, and the open subsets of $\mathfrak{V}$ are the unions of the clopen sets. Each element F in $\mathfrak{C}$ comes from $\mathfrak{A}$ and is therefore a subset of $\mathfrak{U}$; in fact, F has the form $F = \bigcup_i F_i$, where for each index i, the component $F_i = F \cap U_i$ belongs to $\mathfrak{A}_i$ and is therefore a clopen subset of $\mathfrak{U}_i$. For every element r in $\mathfrak{U}$, the set

$$Y_r = \{F \in C : r \in F\}$$

is easily seen to be an ultrafilter of elements in $\mathfrak{C}$; in fact, if r belongs to the component space $\mathfrak{U}_i$, then Y_r consists of precisely those elements F

in $\mathfrak{C}$ such that r belongs to the component F_i. Every ultrafilter of this form is a point in the dual space $\mathfrak{V}$ (but there are other points—other ultrafilters—in $\mathfrak{V}$ as well). Notice that distinct points r and s in $\mathfrak{U}$ yield distinct ultrafilters Y_r and Y_s. Indeed, suppose r is in $\mathfrak{U}_i$ and s in $\mathfrak{U}_j$. If $i \neq j$, then the clopen set U_i—which belongs to $\mathfrak{A}_i$ and is therefore in $\mathfrak{D}$, and hence also in $\mathfrak{C}$—belongs to Y_r, while its complement in $\mathfrak{C}$ belongs to Y_s. If $i = j$, then there is a clopen subset of $\mathfrak{U}_i$ that contains r, but not s, because the clopen subsets of $\mathfrak{U}_i$ separate points; and that clopen set (which belongs to $\mathfrak{A}_i$ and is therefore in $\mathfrak{D}$ and in $\mathfrak{C}$) belongs to Y_r, while its complement in $\mathfrak{U}_i$ belongs to Y_s. Notice also that

$$G_{U_i} = \{Y \in V : U_i \in Y\}$$

is a clopen subset of $\mathfrak{V}$ (because U_i belongs to $\mathfrak{D}$ and therefore also to $\mathfrak{C}$).

Define a subset W of $\mathfrak{V}$ by

$$W = \{Y_r : r \in U\}.$$

It is easy to check that W is a dense subset of $\mathfrak{V}$. For the proof, it suffices to show that every non-empty clopen set in $\mathfrak{V}$ has a non-empty intersection with W. An arbitrary non-empty clopen set in $\mathfrak{V}$ has the form G_F for some non-empty set F in $\mathfrak{C}$, by the remarks of the preceding paragraph. It follows from (1) and the definitions of the set W and the ultrafilters Y_r that

$$G_F \cap W = \{Y_r : F \in Y_r\} = \{Y_r : r \in F\}. \tag{2}$$

The set F is not empty, so the intersection in (2) cannot be empty. In fact, it contains the element Y_r for every r in F. Thus, the set W is dense in $\mathfrak{V}$, as claimed. We shall eventually prove that the restriction of $\mathfrak{V}$ to W is an inner subspace of $\mathfrak{V}$ that is homeo-isomorphic to $\mathfrak{U}$. Write

$$W_i = G_{U_i} \cap W = \{Y_r : r \in U_i\}, \tag{3}$$

and observe that for $i \neq j$, the sets W_i and W_j are disjoint, since the sets U_i and U_j are disjoint. The space $\mathfrak{U}$ is the disjoint union of the component spaces $\mathfrak{U}_i$, so obviously W is the disjoint union of the component sets W_i (in $\mathfrak{V}$), by the definitions of W and W_i, and by the observation made earlier that distinct elements r and s in $\mathfrak{U}$ lead to distinct ultrafilters Y_r and Y_s. The immediate goal is to prove that W_i is a compact subset of $\mathfrak{V}$. Assume for a moment that this has been accomplished. It then follows that W, as the disjoint union of compact subsets of $\mathfrak{V}$, is locally compact. A dense, locally compact subspace of a Hausdorff space is necessarily open (see Corollary 1 on p. 400 of [10]), so W must be open in $\mathfrak{V}$. Thus, W_i is the intersection of two open sets in $\mathfrak{V}$, namely G_{U_i} and W, so W_i must also be open in $\mathfrak{V}$. Of course, W_i is also closed in $\mathfrak{V}$, because it is a compact subset of the Hausdorff space $\mathfrak{V}$, so W_i must in fact be clopen.

The proof that W_i is compact is somewhat involved and uses some of the duality theorems that were proved earlier. Because $\mathfrak{A}$ is the internal product of the system of algebras $(\mathfrak{A}_i : i \in I)$, the projection from $\mathfrak{A}$ to the factor algebra $\mathfrak{A}_i$ is the epimorphism φ_i defined by

$$\varphi_i(F) = F \cap U_i = F_i$$

for every set F in $\mathfrak{A}$, where F_i is the component of F in $\mathfrak{A}_i$. Because $\mathfrak{C}$ is a subalgebra of $\mathfrak{A}$, the restriction of φ_i to $\mathfrak{C}$ is a homomorphism from $\mathfrak{C}$ into $\mathfrak{A}_i$. Every element in $\mathfrak{A}_i$ belongs to $\mathfrak{D}$ and therefore also to $\mathfrak{C}$. For each set F in $\mathfrak{A}_i$, we have

$$\varphi_i(F) = F \cap U_i = F,$$

since F is a subset of $\mathfrak{U}_i$. Thus, φ_i maps $\mathfrak{C}$ homomorphically onto $\mathfrak{A}_i$. We shall refer to the restriction of φ_i to $\mathfrak{C}$ by using the same symbol φ_i.

The restriction of φ_i to $\mathfrak{C}$ induces a continuous bounded monomorphism ϑ_i from the dual space of $\mathfrak{A}_i$ to the dual space of $\mathfrak{C}$ that is defined by

$$\vartheta_i(X) = \varphi_i^{-1}(X)$$

for each element X in the dual space of $\mathfrak{A}_i$, by Theorem 2.18. The dual space of $\mathfrak{C}$ is $\mathfrak{V}$, by assumption. The dual space of $\mathfrak{A}_i$—call it $\bar{\mathfrak{U}}_i$—is the second dual of the relational space $\mathfrak{U}_i$, and $\mathfrak{U}_i$ is homeo-isomorphic $\bar{\mathfrak{U}}_i$ via the function δ_i that maps each element r in $\mathfrak{U}_i$ to the ultrafilter

$$X_r = \{F \in A_i : r \in F\},$$

by Theorem 2.11. In particular, the elements in $\bar{\mathfrak{U}}_i$ are precisely the ultrafilters of the form X_r for elements r in $\mathfrak{U}_i$, and distinct elements in $\mathfrak{U}_i$ correspond to distinct ultrafilters. The definition of ϑ_i may therefore be written in the form

$$\vartheta_i(X_r) = \varphi_i^{-1}(X_r)$$

for each element r in $\mathfrak{U}_i$. Observe that for each such r,

$$\varphi_i^{-1}(X_r) = \{F \in C : \varphi_i(F) \in X_r\} = \{F \in C : F \cap U_i \in X_r\}$$
$$= \{F \in C : r \in F \cap U_i\} = \{F \in C : r \in F\} = Y_r,$$

by the definition of the inverse image under φ_i of a set, the definition of φ_i, the definition of X_r, the assumption that r is in $\mathfrak{U}_i$, and the definition of Y_r. Combine these observations to conclude that ϑ_i is the continuous bounded monomorphism from $\bar{\mathfrak{U}}_i$ to $\mathfrak{V}$ that is determined by

$$\vartheta_i(X_r) = Y_r \tag{4}$$

for every r in $\mathfrak{U}_i$.

The set W_i is the image (in $\mathfrak{V}$) under the continuous mapping ϑ_i of the compact set $\bar{U}_i$, by (3) and (4). The continuous image of a compact set is compact, so W_i must be a compact subset of $\mathfrak{V}$. The argument presented earlier now implies that W_i is a clopen subset, and W an open subset, of $\mathfrak{V}$. From this, it is not difficult to see that the subspace topology on W coincides with the union topology that W inherits from the components W_i. Indeed, the open sets in W under the subspace topology are just the subsets of W that are open in $\mathfrak{V}$, because W itself is open in $\mathfrak{V}$. For any subset H of W, write $H_i = H \cap W_i$, and observe that

$$H = H \cap W = H \cap (\textstyle\bigcup_i W_i) = \textstyle\bigcup_i (H \cap W_i) = \textstyle\bigcup_i H_i.$$

If H is open in $\mathfrak{V}$, then H_i is open in $\mathfrak{V}$, because W_i is open in $\mathfrak{V}$; consequently, H is a union of open subsets of the components W_i, so H is open in the union topology, by the definition of that topology. On the other hand, if H is open in the union topology, then each set H_i is open in W_i, and therefore also open in $\mathfrak{V}$, by the definition of the union topology; consequently, H is a union of open sets in $\mathfrak{V}$, so H is open in $\mathfrak{V}$.

The fact that ϑ_i is a continuous bounded monomorphism from $\bar{\mathfrak{U}}_i$ into $\mathfrak{V}$ implies that the image set W_i is an inner subuniverse of $\mathfrak{V}$, by Lemma 1.13, and also that ϑ_i is algebraically an isomorphism from $\bar{\mathfrak{U}}_i$ to the corresponding inner substructure that is the restriction of $\mathfrak{V}$ to W_i. The set W_i is closed in $\mathfrak{V}$, so the restriction of $\mathfrak{V}$ to W_i is actually a relational space $\mathfrak{W}_i$ that is an inner subspace of $\mathfrak{V}$, by Lemma 2.25. It is clear that ϑ_i is continuous with respect to the topology on $\mathfrak{W}_i$. Indeed, W_i is open in $\mathfrak{V}$, so every open subset of $\mathfrak{W}_i$

is open in $\mathfrak{V}$, and therefore the inverse image under ϑ_i of every open subset of $\mathfrak{W}_i$ must be open in $\bar{\mathfrak{U}}_i$ (because ϑ_i is continuous with respect to the topology on $\mathfrak{V}$). A continuous bijection from a compact space to a Hausdorff space is necessarily a homeomorphism (see Lemma 5 on p. 316 of [10]), so ϑ_i is a homeo-isomorphism from $\bar{\mathfrak{U}}_i$ to $\mathfrak{W}_i$.

Write $\mathfrak{W}$ for the restriction of $\mathfrak{V}$ to the set W. It is not difficult to check that $\mathfrak{V}$ is a compactification of $\mathfrak{W}$ in the sense of Definition 2.48. First of all, the topology on $\mathfrak{W}$ is, by definition, the topology inherited from $\mathfrak{V}$. Second, we have already seen that the universe of $\mathfrak{W}$ is a dense subset of $\mathfrak{V}$. Third, $\mathfrak{W}$ is algebraically an inner substructure of $\mathfrak{V}$. To see this, consider the case of a ternary relation R. Let r and s be any elements in $\mathfrak{V}$, and t any element in $\mathfrak{W}$, and suppose that $R(r, s, t)$ holds in $\mathfrak{V}$. The universe of $\mathfrak{W}$ is the union of the universes of the spaces $\mathfrak{W}_i$, so the element t must belong to $\mathfrak{W}_i$ for some i. It was shown in the preceding paragraph that $\mathfrak{W}_i$ is an inner subspace of $\mathfrak{V}$, so the elements r and s must belong to $\mathfrak{W}_i$, and therefore these elements must also belong to $\mathfrak{W}$.

In a similar way, we show that $\mathfrak{W}$ is the union, in the sense of Definition 2.43, of the disjoint system of relational spaces $(\mathfrak{W}_i : i \in I)$. First of all, we have already seen that the subspace topology on $\mathfrak{W}$ coincides with the union topology inherited from the given system of spaces. Second, to check that $\mathfrak{W}$ is algebraically the union of the relational structures $\mathfrak{W}_i$, consider as an example the case of a ternary relation R. If $R(r, s, t)$ holds in $\mathfrak{W}$, then this relationship must also hold in $\mathfrak{V}$, because $\mathfrak{W}$ is a restriction of $\mathfrak{V}$. The element t belongs to $\mathfrak{W}_i$ for some i, and $\mathfrak{W}_i$ is an inner subspace of $\mathfrak{V}$, so r and s must be in $\mathfrak{W}_i$. Consequently, $R(r, s, t)$ holds in $\mathfrak{W}_i$, because $\mathfrak{W}_i$ is a restriction of $\mathfrak{V}$. Conversely, if $R(r, s, t)$ holds in $\mathfrak{W}_i$, then this relationship also holds in $\mathfrak{V}$, because $\mathfrak{W}_i$ is a restriction of $\mathfrak{V}$. The elements r, s, and t clearly belong to $\mathfrak{W}$, which is a restriction of $\mathfrak{V}$, so $R(r, s, t)$ must hold in $\mathfrak{W}$. Conclusion: the relation R in $\mathfrak{W}$ is the (disjoint) union of the corresponding relations in the spaces $\mathfrak{W}_i$, so $\mathfrak{W}$ is the union of the given system, as claimed.

The composition $\varrho_i = \vartheta_i \circ \delta_i$ of the homeo-isomorphism δ_i from $\mathfrak{U}_i$ to $\bar{\mathfrak{U}}_i$ and the homeo-isomorphism ϑ_i from $\bar{\mathfrak{U}}_i$ to $\mathfrak{W}_i$ is a homeo-isomorphism from $\mathfrak{U}_i$ to $\mathfrak{W}_i$. The union of these compositions is the bijection ϱ from $\mathfrak{U}$ to $\mathfrak{W}$ that is defined by

$$\varrho(r) = \varrho_i(r) = \vartheta_i(\delta_i(r)) = \vartheta_i(X_r) = Y_r \tag{5}$$

whenever r is an element in $\mathfrak{U}$ that belongs to $\mathfrak{U}_i$. It is not difficult to check that this disjoint union of homeo-isomorphisms is itself a homeo-isomorphism. To check that ϱ is a homeomorphism, consider any open subset $F = \bigcup_i F_i$ of $\mathfrak{U}$. The component sets F_i must be open in $\mathfrak{U}_i$, by the definition of the union topology, and each mapping ϱ_i is a homeo-isomorphism, so each set $\varrho_i(F_i)$ must be open in $\mathfrak{W}_i$. Consequently, the union of these sets is open in $\mathfrak{W}$ (because the topology on $\mathfrak{W}$ coincides with the union topology). That union is just $\varrho(F)$, since the image of F under ϱ is the set

$$\varrho(F) = \varrho(\textstyle\bigcup_i F_i) = \bigcup_i \varrho(F_i) = \bigcup_i \varrho_i(F_i),$$

so the image under ϱ of an open set in $\mathfrak{U}$ is an open set in $\mathfrak{W}$. A completely analogous argument shows that the inverse image under ϱ of an open set in $\mathfrak{W}$ is an open set in $\mathfrak{U}$. Thus, ϱ is a homeomorphism, as claimed.

To check that ϱ isomorphically preserves the fundamental relations, consider the case of a ternary relation R. Let r, s, and t be elements in $\mathfrak{U}$. If $R(r, s, t)$ holds in $\mathfrak{U}$, then this relationship must hold in $\mathfrak{U}_i$ for some index i, because $\mathfrak{U}$ is the union of the component spaces $\mathfrak{U}_i$. Since ϱ_i is a homeo-isomorphism from $\mathfrak{U}_i$ to $\mathfrak{W}_i$, the relationship

$$R(\varrho_i(r), \varrho_i(s), \varrho_i(t))$$

must hold in $\mathfrak{W}_i$. Consequently, $R(\varrho(r), \varrho(s), \varrho(t))$ must hold in $\mathfrak{W}_i$ and therefore also in $\mathfrak{W}$, by the definition of ϱ and the fact that $\mathfrak{W}$ is the union of the component spaces $\mathfrak{W}_i$. A completely analogous argument shows that if $R(\varrho(r), \varrho(s), \varrho(t))$ holds in $\mathfrak{W}$, then $R(r, s, t)$ holds in $\mathfrak{U}$. Thus, ϱ isomorphically preserves the relation R. Conclusion: ϱ is an isomorphism, and therefore a homeo-isomorphism, from $\mathfrak{U}$ to $\mathfrak{W}$.

The homeo-isomorphism ϱ from $\mathfrak{U}$ to $\mathfrak{W}$ induces an isomorphism $\bar{\varrho}$ from $\mathfrak{Cm}(U)$ to $\mathfrak{Cm}(W)$ that is defined by

$$\bar{\varrho}(F) = \{\varrho(r) : r \in F\}$$

for every subset F of $\mathfrak{U}$, by Corollary 1.7. In view of (5), it is clear that the definition of $\bar{\varrho}$ may be written in the form

$$\varrho(F) = \{Y_r : r \in F\}. \tag{6}$$

The algebra $\mathfrak{C}$ is a subalgebra of $\mathfrak{A}$, by assumption, and $\mathfrak{A}$ is a subalgebra of $\mathfrak{Cm}(U)$, by Theorem 1.32, so $\mathfrak{C}$ is a subalgebra of $\mathfrak{Cm}(U)$.

It therefore makes sense to restrict the isomorphism $\bar{\varrho}$ to $\mathfrak{C}$, and this restriction must map $\mathfrak{C}$ isomorphically to a subalgebra of $\mathfrak{Cm}(W)$.

Consider now the dual algebra of $\mathfrak{V}$—call it $\mathfrak{B}$. This algebra is the second dual of $\mathfrak{C}$, and $\mathfrak{C}$ is isomorphic to its second dual via the canonical isomorphism ζ that maps each set F in $\mathfrak{C}$ to the set G_F defined in (1), by Theorem 2.10. In particular, the elements in $\mathfrak{B}$ are just the sets G_F for F in $\mathfrak{C}$. The algebra $\mathfrak{B}$ is, in turn, isomorphic to its relativization $\mathfrak{B}_0$ via the relativization isomorphism ψ that maps each element in $\mathfrak{B}$ to its intersection with W, by Lemma 2.50 (with $\mathfrak{W}$ as the locally compact relational space, and $\mathfrak{V}$ as the compactification of $\mathfrak{W}$). In this connection, recall from the proof of Lemma 2.50 that ψ is the restriction of the relativization homomorphism from $\mathfrak{Cm}(V)$ to $\mathfrak{Cm}(W)$ that maps each subset of $\mathfrak{V}$ to its intersection with the set W. Composing ψ with ζ, and using (2), we arrive at

$$(\psi \circ \zeta)(F) = \psi(\zeta(F)) = \psi(G_F) = G_F \cap W = \{Y_r : r \in F\} \qquad (7)$$

for each set F in $\mathfrak{C}$. Compare (6) with (7) to conclude that

$$\bar{\varrho} = \psi \circ \zeta$$

(where the left side of this equation actually denotes the restriction of $\bar{\varrho}$ to $\mathfrak{C}$). Thus, the restriction of $\bar{\varrho}$ maps $\mathfrak{C}$ isomorphically to the algebra $\mathfrak{B}_0$ that is the relativization to W of the algebra $\mathfrak{B}$, which in turn is the dual of $\mathfrak{V}$.

Here is a summary of what has been accomplished. First, the locally compact union space $\mathfrak{U}$ has been mapped homeo-isomorphically by ϱ to a space $\mathfrak{W}$ of which $\mathfrak{V}$ is a compactification. Second, the subalgebra $\mathfrak{C}$ of $\mathfrak{A}$ has been mapped isomorphically by $\bar{\varrho}$—the mapping induced on $\mathfrak{C}$ by ϱ—to the relativization of the dual algebra $\mathfrak{B}$ (of $\mathfrak{V}$) to the set W. If we now use the homeo-isomorphism ϱ to identify $\mathfrak{U}$ with $\mathfrak{W}$, and we use the induced isomorphism $\bar{\varrho}$ to identify $\mathfrak{C}$ with the relativization of $\mathfrak{B}$ to W, then we arrive at the desired goal: $\mathfrak{V}$ is a compactification of the union space $\mathfrak{U}$, and the relativization to U of the dual algebra of $\mathfrak{V}$ is just $\mathfrak{C}$.

The technical tool for carrying out this identification is a version of the general algebraic Exchange Principle that applies to structures such as $\mathfrak{U}$. The elements in $\mathfrak{V}$ that come from $\mathfrak{W}$ are replaced by the corresponding elements from $\mathfrak{U}$ (under the correspondence that is the inverse of ϱ), the remaining elements in $\mathfrak{V}$ being modified if necessary so that they do not occur in $\mathfrak{U}$. Once this is accomplished, the function ϱ becomes the identity function on $\mathfrak{U}$, and therefore the mapping $\bar{\varrho}$ on $\mathfrak{C}$ induced by ϱ becomes the identity function on $\mathfrak{C}$. Consequently, $\mathfrak{C}$ coincides with $\mathfrak{B}_0$, which is the relativization of $\mathfrak{B}$ to W. $\square$

Consider the function that maps each compactification $\mathfrak{V}$ of the union space $\mathfrak{U}$ to the isomorphic copy of the dual algebra of $\mathfrak{V}$ that is obtained by relativizing the dual algebra to the set U. This function maps the class of compactifications of $\mathfrak{U}$ onto the class of algebras intermediate between $\mathfrak{D}$ and $\mathfrak{A}$, by Lemmas 2.50 and 2.53. In general, this function is not one-to-one, as Lemma 2.52 makes clear; distinct compactifications of $\mathfrak{U}$ may be mapped to the same intermediate algebra. Such compactifications do not differ from one another in any material way, and it is natural to identify them by grouping them together in one class. Motivated by these considerations, we define two compactifications of $\mathfrak{U}$ to be *equivalent* if there is a homeo-isomorphism from one compactification to the other that is the identity function on $\mathfrak{U}$. It is easy to check that the relation defined in this way is an equivalence relation on the class of compactifications of $\mathfrak{U}$. Equivalent compactifications have dual algebras that are isomorphic via a function that is the identity on the universes of the factor algebras $\mathfrak{A}_i$, and the isomorphic copies of these dual algebras obtained by relativization to U are in fact equal, by Lemma 2.52. Thus, one may speak with some justification of the dual algebra of the equivalence class. The correspondence that maps each equivalence class of compactifications of $\mathfrak{U}$ to the relativization of its dual algebra is a well-defined bijection from the class of equivalence classes of compactifications of $\mathfrak{U}$ to the set of subalgebras of $\mathfrak{A}$ that include $\mathfrak{D}$. It turns out that this bijection is actually a lattice isomorphism.

The set of algebras between $\mathfrak{D}$ and $\mathfrak{A}$ is partially ordered by the relation of being a subalgebra, and under this partial ordering the set becomes a complete lattice with zero $\mathfrak{D}$ and unit $\mathfrak{A}$. The partial ordering on the class of equivalence classes of compactifications of $\mathfrak{U}$ is more complicated to describe, but it is equally natural. Define a binary relation $\leq$ on the class of compactifications of $\mathfrak{U}$ by writing $\mathfrak{W} \leq \mathfrak{V}$ just in case there is a continuous bounded epimorphism from $\mathfrak{V}$ to $\mathfrak{W}$ that is the identity function on $\mathfrak{U}$. This relation is preserved by the relation of equivalence in the following sense: if compactifications $\mathfrak{V}_1$ and $\mathfrak{V}_2$ are equivalent, and if compactifications $\mathfrak{W}_1$ and $\mathfrak{W}_2$ are also equivalent, then

$$\mathfrak{W}_1 \leq \mathfrak{V}_1 \qquad \text{if and only if} \qquad \mathfrak{W}_2 \leq \mathfrak{V}_2.$$

For the proof, suppose δ is a homeo-isomorphism from $\mathfrak{V}_1$ to $\mathfrak{V}_2$ that is the identity function on $\mathfrak{U}$, and ϱ a homeo-isomorphism from $\mathfrak{W}_1$ to $\mathfrak{W}_2$ that is the identity function on $\mathfrak{U}$. If ϑ is a continuous bounded epimorphism from $\mathfrak{V}_1$ to $\mathfrak{W}_1$ that is the identity function on $\mathfrak{U}$, then the composition $\varrho \circ \vartheta \circ \delta^{-1}$ is a continuous bounded epimorphism from $\mathfrak{V}_2$ to $\mathfrak{W}_2$ that is the identity function on $\mathfrak{U}$ (see the diagram below). This shows that $\mathfrak{W}_1 \leq \mathfrak{V}_1$ implies $\mathfrak{W}_2 \leq \mathfrak{V}_2$. The reverse implication is established by a symmetric argument.

$$
\begin{array}{ccc}
\mathfrak{V}_1 & \xrightarrow{\ \delta\ } & \mathfrak{V}_2 \\
\vartheta \downarrow & & \downarrow \\
\mathfrak{W}_1 & \xrightarrow[\varrho]{} & \mathfrak{W}_2
\end{array}
$$

We shall say of two compactifications $\mathfrak{V}$ and $\mathfrak{W}$ of $\mathfrak{U}$ that the equivalence class of $\mathfrak{W}$ is *less than or equal to* the equivalence class of $\mathfrak{V}$ if $\mathfrak{W} \leq \mathfrak{V}$. The remarks in the preceding paragraph imply that the relation on equivalence classes defined in this manner is well defined in the sense that it does not depend on the particular choice of the representatives of the equivalence classes involved. One can prove without difficulty that the relation is a partial ordering on the class of equivalence classes. For instance, to prove that the relation is anti-symmetric, assume that the equivalence class of $\mathfrak{W}$ is less than or equal to the equivalence class of $\mathfrak{V}$, and vice versa. It must be shown that the two equivalence classes are equal. The assumption implies that $\mathfrak{V} \leq \mathfrak{W}$ and $\mathfrak{W} \leq \mathfrak{V}$. Consequently, there is a continuous bounded epimorphism ϑ from $\mathfrak{V}$ to $\mathfrak{W}$ and a continuous bounded epimorphism δ from $\mathfrak{W}$ to $\mathfrak{V}$ such that both mappings are the identity function on $\mathfrak{U}$, by the definition of $\leq$. The composition $\delta \circ \vartheta$ is therefore a continuous bounded epimorphism from $\mathfrak{V}$ to $\mathfrak{V}$ that is the identity function on $\mathfrak{U}$. The identity function on $\mathfrak{V}$ is also a continuous bounded epimorphism from $\mathfrak{V}$ to $\mathfrak{V}$ that is the identity function on $\mathfrak{U}$. The universe of $\mathfrak{U}$ is a dense subset of $\mathfrak{V}$, by the assumption that $\mathfrak{V}$ is a compactification of $\mathfrak{U}$. Two continuous functions from a topological space to a Hausdorff space that agree on a dense subset must agree on the entire space (see Corollary 2 on p. 315 of [10]), so the composition $\delta \circ \vartheta$ must be the identity function on $\mathfrak{V}$. A similar argument shows that $\vartheta \circ \delta$ is the identity function on $\mathfrak{W}$. Consequently, ϑ is a bijection from $\mathfrak{V}$ to $\mathfrak{W}$ and δ is its inverse. A bounded epimorphism that is a bijection is an isomorphism, by the remark following Definition 1.8, so ϑ is a continuous isomorphism with a continuous inverse δ. Conclusion: ϑ is a homeo-isomorphism that is the identity function on $\mathfrak{U}$. This implies that the compactifications $\mathfrak{V}$ and $\mathfrak{W}$ are equivalent, and therefore their equivalence classes are equal, as desired.

We have seen that the class of equivalence classes of compactifications of $\mathfrak{U}$ is partially ordered by the less-than-or-equal-to relation defined in the preceding paragraph, and the class of subalgebras of $\mathfrak{A}$ that include $\mathfrak{D}$ is partially ordered by the relation of being a subalgebra. We have also seen that the function ζ mapping each equivalence class of compactifications to the relativization to U of the dual algebra of the equivalence class is a bijection from the class of equivalence classes of compactifications of $\mathfrak{U}$ to the class of subalgebras of $\mathfrak{A}$ that include $\mathfrak{D}$. We now show that ζ preserves the partial ordering in a strong sense. Consider compactifications $\mathfrak{V}$ and $\mathfrak{W}$ of $\mathfrak{U}$, and let $\mathfrak{B}$ and $\mathfrak{C}$ be their respective dual algebras. Write $\mathfrak{B}_0$ and $\mathfrak{C}_0$ for the relativizations of $\mathfrak{B}$ and $\mathfrak{C}$ to U. The function ζ maps the equivalence class of $\mathfrak{V}$ to the subalgebra $\mathfrak{B}_0$, and the equivalence class of $\mathfrak{W}$ to the subalgebra $\mathfrak{C}_0$. Since the equivalence class of $\mathfrak{W}$ is less than or equal to the equivalence class of $\mathfrak{V}$ just in case $\mathfrak{W} \leq \mathfrak{V}$, it suffices to prove that

$$\mathfrak{W} \leq \mathfrak{V} \qquad \text{if and only if} \qquad \mathfrak{C}_0 \subseteq \mathfrak{B}_0.$$

If $\mathfrak{W} \leq \mathfrak{V}$, then there must be a continuous bounded epimorphism ϑ from $\mathfrak{V}$ to $\mathfrak{W}$ that is the identity function on $\mathfrak{U}$, by the definition of the relation $\leq$. Apply Lemma 2.51 to conclude that the relativization of $\mathfrak{C}$ to U is a subalgebra of the relativization of $\mathfrak{B}$ to U, that is to say, $\mathfrak{C}_0$ is a subalgebra of $\mathfrak{B}_0$. Conversely, if $\mathfrak{C}_0$ is a subalgebra of $\mathfrak{B}_0$, then there must be a continuous bounded epimorphism from $\mathfrak{V}$ to $\mathfrak{W}$ that is the identity function on U, again by Lemma 2.51, so $\mathfrak{W} \leq \mathfrak{V}$. The following theorem summarizes what has been proved.

Theorem 2.54. *Let $\mathfrak{U}$ be the union of a disjoint system $(\mathfrak{U}_i : i \in I)$ of relational spaces. For each index i, let $\mathfrak{A}_i$ be the dual algebra of $\mathfrak{U}_i$, and let $\mathfrak{A}$ be the internal product, and $\mathfrak{D}$ the weak internal product, of the system $(\mathfrak{A}_i : i \in I)$ of algebras. Equivalent compactifications of $\mathfrak{U}$ have, up to isomorphism, the same dual algebra, and that dual algebra is isomorphic via relativization to a subalgebra of $\mathfrak{A}$ that includes $\mathfrak{D}$. The function that maps each equivalence class of compactifications of $\mathfrak{U}$ to the corresponding subalgebra of $\mathfrak{A}$ that includes $\mathfrak{D}$ is a lattice isomorphism from the lattice of equivalence classes of compactifications of $\mathfrak{U}$ to the lattice of subalgebras of $\mathfrak{A}$ that include $\mathfrak{D}$.*

2.12 Duality for Infinite Direct Products

The internal product $\mathfrak{A}$ of the system of dual algebras $(\mathfrak{A}_i : i \in I)$ of a given disjoint system $(\mathfrak{U}_i : i \in I)$ of relational spaces is the maximum element in the lattice of subalgebras of $\mathfrak{A}$ that include the weak internal product. The equivalence class of the dual space of $\mathfrak{A}$ must therefore be the maximum element in the lattice of equivalence classes of compactifications of the union space $\mathfrak{U}$, by Theorem 2.54. It is natural to look for a topological characterization of this maximum compactification. The maximum compactification of an arbitrary locally compact Hausdorff space U is the Stone-Čech compactification. This space—call it V—is characterized by the property that every continuous mapping from U into a compact Hausdorff space W can be extended (in a unique way) to a continuous mapping from V into W. These considerations motivate the following definition.

Definition 2.55. A *Stone-Čech compactification* of a locally compact relational space $\mathfrak{U}$ is defined to be a compactification $\mathfrak{V}$ of $\mathfrak{U}$ with the property that every continuous bounded homomorphism from $\mathfrak{U}$ into a relational space $\mathfrak{W}$ can be extended to a continuous bounded homomorphism from $\mathfrak{V}$ into $\mathfrak{W}$.   □

Observe that if $\mathfrak{V}$ is a Stone-Čech compactification of a locally compact relational space $\mathfrak{U}$, then the continuous bounded homomorphism on $\mathfrak{V}$ that extends a given continuous bounded homomorphism from $\mathfrak{U}$ to a relational space $\mathfrak{W}$ must be unique. The reason is that the universe of $\mathfrak{U}$ is a dense subset of $\mathfrak{V}$, by Definition 2.48, and two continuous functions that agree on a dense subset of a space agree everywhere on the space. Observe also that if such a compactification $\mathfrak{V}$ exists, then the equivalence class of $\mathfrak{V}$ must be the maximum element in the lattice of equivalence classes of compactifications of $\mathfrak{U}$, by the next lemma.

Lemma 2.56. *If $\mathfrak{V}$ is a Stone-Čech compactification of a locally compact relational space $\mathfrak{U}$, then every compactification of $\mathfrak{U}$ is a continuous bounded homomorphic image of $\mathfrak{V}$ via a mapping that is the identity function on $\mathfrak{U}$.*

Proof. Suppose $\mathfrak{W}$ is a compactification of $\mathfrak{U}$. The space $\mathfrak{U}$ is then an inner subspace of $\mathfrak{W}$ in the sense that it is algebraically an inner substructure of $\mathfrak{W}$ and topologically a subspace of $\mathfrak{W}$ (see Definition 2.48). The identity function ϑ on $\mathfrak{U}$ is therefore a continuous bounded homomorphism from $\mathfrak{U}$ to $\mathfrak{W}$. In more detail, ϑ is a bounded monomorphism from $\mathfrak{U}$ to $\mathfrak{W}$, by Corollary 1.14; and if H is an open subset of $\mathfrak{W}$, then the inverse image of H under ϑ is the set

$$\vartheta^{-1}(H) = \{r \in U : \vartheta(r) \in H\} = \{r \in U : r \in H\} = H \cap U,$$

which is an open subset of $\mathfrak{U}$, by the definition of the subspace topology; consequently, ϑ is continuous. Apply Definition 2.55 and the assumption that $\mathfrak{V}$ is a Stone-Čech compactification of $\mathfrak{U}$ to conclude that there is a continuous bounded homomorphism δ from $\mathfrak{V}$ to $\mathfrak{W}$ that extends ϑ. A continuous mapping from a compact space to a Hausdorff space maps closed sets to closed sets (see Corollary 1 on p. 315 of [10]), so the image set

$$\delta(V) = \{\delta(r) : r \in V\}$$

must be closed in $\mathfrak{W}$. This image set includes the set U, because δ is an extension of ϑ, and ϑ is the identity function on $\mathfrak{U}$. Since U is a dense subset of $\mathfrak{W}$, the closure of U in $\mathfrak{W}$ must be W. It follows that

$$W = U^- \subseteq \delta(V)^- = \delta(V),$$

so δ maps $\mathfrak{V}$ onto $\mathfrak{W}$. Conclusion: $\mathfrak{W}$ is the image of $\mathfrak{V}$ under a continuous bounded homomorphism δ that is the identity function on $\mathfrak{U}$. $\square$

Corollary 2.57. *A Stone-Čech compactification of a locally compact relational space $\mathfrak{U}$, if it exists, is unique up to homeo-isomorphisms that are the identity function on $\mathfrak{U}$.*

Proof. Suppose $\mathfrak{V}_1$ and $\mathfrak{V}_2$ are Stone-Čech compactifications of a locally compact relational space $\mathfrak{U}$. There is a continuous bounded epimorphism δ_1 from $\mathfrak{V}_1$ to $\mathfrak{V}_2$ that is the identity function on $\mathfrak{U}$, and also a continuous bounded epimorphism δ_2 from $\mathfrak{V}_2$ into $\mathfrak{V}_1$ that is the identity function on $\mathfrak{U}$, by Lemma 2.56. Just as in the argument given before Theorem 2.54 (showing that the relation $\leq$ is antisymmetric), the compositions $\delta_1 \circ \delta_2$ and $\delta_2 \circ \delta_1$ must be the identity functions on the spaces $\mathfrak{V}_2$ and $\mathfrak{V}_1$ respectively, so δ_1 and δ_2 are bijections and inverses of one another; consequently, δ_1 is a homeo-isomorphism from $\mathfrak{V}_1$ to $\mathfrak{V}_2$ that is the identity function on $\mathfrak{U}$. $\square$

The preceding corollary justifies speaking about *the* Stone-Čech compactification of a locally compact relational space $\mathfrak{U}$. The next theorem establishes the existence of the Stone-Čech compactification in the case in which we are interested, and it simultaneously shows that this compactification is essentially the dual space of the direct product.

Theorem 2.58. *Let $\mathfrak{U}$ be the union of a disjoint system $(\mathfrak{U}_i : i \in I)$ of relational spaces, and for each index i, let $\mathfrak{A}_i$ be the dual algebra of $\mathfrak{U}_i$. The Stone-Čech compactification of $\mathfrak{U}$ exists, and its dual algebra is isomorphic to the internal product of the system $(\mathfrak{A}_i : i \in I)$ via the relativization function.*

Proof. We begin by introducing some notation and making some preliminary observations. Let $\mathfrak{A}$ be the internal product of the system $(\mathfrak{A}_i : i \in I)$ of dual algebras of the given disjoint system $(\mathfrak{U}_i : i \in I)$ of relational spaces. The definition of the internal product implies that the elements in $\mathfrak{A}$ are the subsets of the union space $\mathfrak{U}$ that can be written in the form $F = \bigcup_i F_i$, where F_i is an element in $\mathfrak{A}_i$ and therefore a clopen subset of $\mathfrak{U}_i$. The clopen subsets of the space $\mathfrak{U}$ are also the unions of the clopen subsets of the component spaces $\mathfrak{U}_i$, by the remarks at the beginning of Section 2.10 concerning union spaces. It follows that the elements in $\mathfrak{A}$ are precisely the clopen subsets of $\mathfrak{U}$.

Observe, as in the proof of Theorem 2.47, that for each index i, the dual algebra $\mathfrak{A}_i$ is, by its very construction, a subalgebra of the complex algebra $\mathfrak{Cm}(U_i)$. The internal product of the system of dual algebras is therefore a subalgebra of the internal product of the system of complex algebras

$$(\mathfrak{Cm}(U_i) : i \in I).$$

The first product is equal to $\mathfrak{A}$, by assumption, and the second product is equal to the complex algebra $\mathfrak{Cm}(U)$, by Theorem 1.32, so $\mathfrak{A}$ is a subalgebra of $\mathfrak{Cm}(U)$. Conclusion: $\mathfrak{A}$ is the subalgebra of $\mathfrak{Cm}(U)$ whose universe is the set clopen subsets of $\mathfrak{U}$, by the previous observations.

Let $\mathfrak{V}$ be the maximum compactification of the union space $\mathfrak{U}$, which exists by Theorem 2.54 and the fact that $\mathfrak{A}$ has a maximum subalgebra, namely itself. The dual algebra of $\mathfrak{V}$—call it $\mathfrak{B}$—is isomorphic to $\mathfrak{A}$ via the relativization function ψ that is defined by

$$\psi(G) = G \cap U$$

for every element G in $\mathfrak{B}$, by Theorem 2.54. The inverse function ψ^{-1} is therefore the isomorphism from $\mathfrak{A}$ to $\mathfrak{B}$ that is defined by

$$\psi^{-1}(F) = G \qquad \text{if and only if} \qquad F = G \cap U, \tag{1}$$

for every element F in $\mathfrak{A}$.

Consider now an arbitrary continuous bounded homomorphism ϑ from $\mathfrak{U}$ into a relational space $\mathfrak{W}$. It is to be shown that ϑ can be extended to a continuous bounded homomorphism δ from $\mathfrak{V}$ to $\mathfrak{W}$. As a bounded homomorphism, ϑ has an algebraic dual, by Theorem 1.9, namely the complete homomorphism φ from $\mathfrak{Cm}(W)$ to $\mathfrak{Cm}(U)$ that is defined by

$$\varphi(H) = \vartheta^{-1}(H) = \{s \in U : \vartheta(s) \in H\} \tag{2}$$

for every element H in $\mathfrak{Cm}(W)$, that is to say, for every subset H of $\mathfrak{W}$. The assumed continuity of ϑ implies that the inverse image under ϑ of every clopen subset of $\mathfrak{W}$ is a clopen subset of $\mathfrak{U}$. Thus, the algebraic dual φ defined in (2) maps clopen subsets of $\mathfrak{W}$ to clopen subsets of $\mathfrak{U}$. The (topological) dual algebra of the relational space $\mathfrak{W}$ is the subalgebra $\mathfrak{C}$ of $\mathfrak{Cm}(W)$ whose universe is the set of all clopen subsets of $\mathfrak{W}$, and the internal product $\mathfrak{A}$ is, by the observations made above, the subalgebra of $\mathfrak{Cm}(U)$ whose universe is the set of all clopen subsets of $\mathfrak{U}$. It follows that the restriction of φ to $\mathfrak{C}$—for which we shall also write φ—is a homomorphism from $\mathfrak{C}$ into $\mathfrak{A}$.

The composition of the homomorphism φ from $\mathfrak{C}$ to $\mathfrak{A}$ with the isomorphism ψ^{-1} from $\mathfrak{A}$ to $\mathfrak{B}$ is a homomorphism ϱ from $\mathfrak{C}$ into $\mathfrak{B}$ that is determined by

$$\varrho(H) = \psi^{-1}(\varphi(H))$$

for every element H in $\mathfrak{C}$. In view of (1) and the definition of ϱ, this means that

$$\varrho(H) = G \quad \text{if and only if} \quad \varphi(H) = G \cap U. \tag{3}$$

Because $\mathfrak{C}$ is the dual algebra of the relational space $\mathfrak{W}$, and $\mathfrak{B}$ is the dual algebra of the relational space $\mathfrak{V}$, and ϱ is a homomorphism from $\mathfrak{C}$ into $\mathfrak{B}$, the topological duality theorem for homomorphisms in the form of Corollary 2.19 (with $\mathfrak{U}$, $\mathfrak{A}$, and φ replaced by $\mathfrak{W}$, $\mathfrak{C}$, and ϱ respectively) may be applied to the homomorphism ϱ to obtain a dual continuous bounded homomorphism δ from $\mathfrak{V}$ to $\mathfrak{W}$ that is defined by

$$\delta(s) = r \quad \text{if and only if} \quad \varrho^{-1}(Y_s) = X_r,$$

where

$$X_r = \{H \in C : r \in H\} \quad \text{and} \quad Y_s = \{G \in B : s \in G\}.$$

It remains to show that δ is an extension of ϑ.

The remarks following Corollary 2.19 (with ϱ and H in place of φ and F) imply that the definition of the dual mapping δ may equivalently be written in the form

$$\delta(s) \in H \qquad \text{if and only if} \qquad s \in \varrho(H) \tag{4}$$

for all elements s in $\mathfrak{V}$ and H in $\mathfrak{C}$. Fix an element H in $\mathfrak{C}$, that is to say, fix a clopen subset of $\mathfrak{W}$, and write $\varrho(H) = G$. For any element s in $\mathfrak{U}$, we have

$$
\begin{aligned}
s \in \varrho(H) \qquad &\text{if and only if} \qquad s \in G, \\
&\text{if and only if} \qquad s \in G \cap U, \\
&\text{if and only if} \qquad s \in \varphi(H), \\
&\text{if and only if} \qquad s \in \vartheta^{-1}(H), \\
&\text{if and only if} \qquad \vartheta(s) \in H,
\end{aligned}
$$

by the definition of the set G, the assumption that s is in $\mathfrak{U}$, the equivalence in (3) and the assumption that $\varrho(H) = G$, the definition of φ in (2), and the definition of the inverse image under ϑ of the set H. The preceding equivalences show that

$$\vartheta(s) \in H \qquad \text{if and only if} \qquad s \in \varrho(H).$$

Combine this with (4) to see that if s belongs to $\mathfrak{U}$, then

$$\delta(s) \in H \qquad \text{if and only if} \qquad \vartheta(s) \in H \tag{9}$$

for every clopen subset H of $\mathfrak{W}$. The clopen subsets of $\mathfrak{W}$ separate points, so the equivalence in (9) can only hold if $\delta(s) = \vartheta(s)$. Thus, δ agrees with ϑ on elements in $\mathfrak{U}$. Conclusion: δ is a continuous bounded homomorphism from $\mathfrak{V}$ to $\mathfrak{W}$ that extends ϑ. $\square$

The preceding theorem does not say that the topology on the Stone-Čech compactification $\mathfrak{V}$ of the union space $\mathfrak{U}$ is the Stone-Čech compactification of the topology on $\mathfrak{U}$ in the standard sense that this term is used in topology, namely that every continuous function from (the universe of) $\mathfrak{U}$ into an arbitrary compact Hausdorff space W can be extended to a continuous function from (the universe of) $\mathfrak{V}$ into W. The theorem only says that condition set forth in Definition 2.55 is satisfied, namely every continuous bounded homomorphism from $\mathfrak{U}$

into a *relational space* $\mathfrak{W}$ can be extended to a continuous bounded homomorphism from $\mathfrak{V}$ to $\mathfrak{W}$. In particular, the topology on $\mathfrak{W}$ is that of a Boolean space, not that of an arbitrary compact Hausdorff space. Showing that the topology on $\mathfrak{V}$ really is the standard Stone-Čech topology requires a separate argument.

Theorem 2.59. *If $\mathfrak{V}$ is the Stone-Čech compactification of the union $\mathfrak{U}$ of a disjoint system of relational spaces, then the topology on $\mathfrak{V}$ is the standard Stone-Čech compactification of the topology on $\mathfrak{U}$.*

Proof. Suppose $\mathfrak{U}$ is the union of a disjoint system

$$(\mathfrak{U}_i : i \in I) \tag{1}$$

of relational spaces. Let

$$(\mathfrak{A}_i : i \in I) \tag{2}$$

be the system of dual algebras corresponding to (1), and let $\mathfrak{A}$ be the internal product of the system of algebras in (2). The dual algebra of the Stone-Čech compactification $\mathfrak{V}$ (of $\mathfrak{U}$)—call it $\mathfrak{B}$—is isomorphic to $\mathfrak{A}$ via the relativization function ψ that maps each set H in $\mathfrak{B}$ to the intersection $H \cap U$, by Theorem 2.58. Essential use of this fact is needed to establish a preliminary observation, namely that two disjoint closed subsets of the union space $\mathfrak{U}$ are always separated by a clopen subset of $\mathfrak{V}$.

For the proof, consider two closed subsets F_1 and F_2 of $\mathfrak{U}$ that are disjoint. For each index i, the intersections

$$F_1 \cap U_i \quad \text{and} \quad F_2 \cap U_i \tag{3}$$

are disjoint closed subsets of $\mathfrak{U}_i$, because $\mathfrak{U}_i$ is topologically a subspace of $\mathfrak{U}$, by Lemma 2.44; and therefore the two sets in (3) are compact, because closed subsets of a compact space are compact. A rather straightforward compactness argument, using the fact that the topology on $\mathfrak{U}_i$ is Boolean, produces a clopen subset G_i of $\mathfrak{U}_i$ that separates the sets in (3) in the sense that

$$F_1 \cap U_i \subseteq G_i \quad \text{and} \quad F_2 \cap G_i = \varnothing. \tag{4}$$

In more detail, in a compact Hausdorff space, any two closed sets can be separated by an open set that includes the first closed set and is disjoint from the second (see Exercise 33 on p. 280 of [10]). If the space is Boolean, then the separating open set is the union of a system of clopen sets, so the compactness of the first closed set implies that there is a finite subsystem of the clopen sets whose union includes the first closed set, and that union remains disjoint from the second closed set. The desired conclusion now follows from the observation that a union of finitely many clopen sets is clopen.

For each index i, the clopen set G_i belongs to the dual algebra $\mathfrak{A}_i$, by the definition of the dual algebra, so the system of clopen sets

$$(G_i : i \in I) \tag{5}$$

has a supremum in the internal product $\mathfrak{A}$, by the definition of the internal product. In fact, the supremum in $\mathfrak{A}$ of this system is just the union

$$G = \bigcup_i G_i.$$

Every clopen subset of $\mathfrak{U}_i$ remains a clopen subset of $\mathfrak{V}$, by Lemma 2.49, and obviously the relativization isomorphism ψ maps each such subset of $\mathfrak{U}_i$ to itself. Consequently, the image of each set G_i under ψ is G_i, and therefore

$$\psi^{-1}(G_i) = G_i. \tag{6}$$

Since the inverse isomorphism ψ^{-1} preserves arbitrary suprema, and since the system in (5) has a supremum in $\mathfrak{A}$, the image of this system under ψ^{-1} must have a supremum in $\mathfrak{B}$. The image system is again just (5), by (6); consequently, the system in (5) has a supremum in $\mathfrak{B}$. Apply Lemma 2.40 (with F_i, H, $\mathfrak{A}$, and $\mathfrak{U}$ respectively replaced by G_i, G, $\mathfrak{B}$, and $\mathfrak{V}$) to conclude that the closure G^- is open and hence clopen in the topology of $\mathfrak{V}$, and that G^- is the supremum of (5) in $\mathfrak{B}$, that is to say

$$G^- = \sum_i G_i.$$

An easy computation shows that the clopen set G^- separates the sets F_1 and F_2 in $\mathfrak{V}$. Indeed,

$$F_1 = F_1 \cap U = F_1 \cap \left(\bigcup_i U_i\right) = \bigcup_i (F_1 \cap U_i) \subseteq \bigcup_i G_i \subseteq G^-,$$

and

$$F_2 \cap G^- = F_2 \cap \left(\sum_i G_i\right) = \sum_i (F_2 \cap G_i) = \sum_i \varnothing = \varnothing,$$

by (4) and the distributive law for multiplication over arbitrary sums in $\mathfrak{B}$.

Turn now to the main task of the proof, which is to demonstrate that the topology on $\mathfrak{V}$ is the Stone-Čech compactification of the topology on $\mathfrak{U}$. We use italic letters to refer to topological spaces having no algebraic structure. Consider an arbitrary continuous function ϑ from U (the universe of $\mathfrak{U}$) into a compact Hausdorff space W. It is to be shown that ϑ can be extended to a continuous function δ from V (the universe of $\mathfrak{V}$) into W. Every point in V is completely determined by the clopen sets to which it belongs, since clopen sets separate points in a Boolean space. Thus, if s is a point in V, and if N_s is the set of clopen subsets of V that contain s, then N_s is an ultrafilter in $\mathfrak{B}$, and the intersection of the sets in N_s is just the singleton $\{s\}$. It is natural to define $\delta(s)$ to be the intersection of the class of image sets

$$\{\vartheta(F) : F \in N_s\}.$$

Two problems arise with this approach. First, ϑ is defined only on points in U (not on points in V), so it is necessary to replace the set $\vartheta(F)$ with the set

$$\vartheta(F \cap U) = \{\vartheta(t) : t \in F \cap U\}.$$

Second, $\vartheta(F \cap U)$ may not be closed in W, so there is no assurance that the intersections of all of the sets of this form will be non-empty. The solution is to pass to the closure $\vartheta(F \cap U)^-$.

Given any point s in V, put

$$M_s = \{\vartheta(F \cap U)^- : F \in N_s\}. \tag{7}$$

We shall prove that the intersection of the sets in M_s contains exactly one point. In order to show that the intersection is not empty, it suffices to prove that the sets in M_s have the finite intersection property in the sense that the intersection of finitely many of these sets is always non-empty; the desired conclusion then follows by compactness, because the intersection of a system of closed sets with the finite intersection property is always non-empty in a compact space (see p. 271 of [10]). The finite intersection property is a direct consequence of two observations. The first is that the sets in M_s are not empty. For the proof, consider a clopen set F in N_s, and notice that F is not empty because it contains s. The set U is a dense subset of V, because V is a compactification of U, so every non-empty open set has a non-empty intersection with U; in particular, F has a non-empty intersection with U. It follows that the image set $\vartheta(F \cap U)$ cannot be empty, so the closure of this image set cannot be empty. The second observation is that the intersection of any finite system of sets in M_s includes a set from M_s and is therefore not empty. For the proof, consider a finite system

$$(F_j : 0 \leq j < n)$$

of sets in N_s. The intersection F of this system belongs to N_s, because N_s is a Boolean filter and is therefore closed under finite intersections. Since F is included in F_j for $j < n$, the set $\vartheta(F \cap U)^-$ must be included in the set $\vartheta(F_j \cap U)^-$ for each j, so that

$$\vartheta(F \cap U)^- \subseteq \bigcap_j \vartheta(F_j \cap U)^-,$$

as desired.

The argument that the intersection of M_s cannot contain two distinct points proceeds by showing that for any two distinct points in W, there is a set in M_s that does not contain at least one of the two points. Let r_1 and r_2 be distinct points in W. Since W is assumed to be a compact Hausdorff space, there must exist open sets H_1 and H_2 in W containing r_1 and r_2 respectively such that the closures H_1^- and H_2^- are disjoint (see Corollary 2 on p. 273 of [10]). The inverse images

$$\vartheta^{-1}(H_1^-) \quad \text{and} \quad \vartheta^{-1}(H_2^-) \tag{8}$$

are then obviously disjoint, and they are closed subsets of U, by the assumed continuity of the mapping ϑ. The preliminary observation at the beginning of the proof implies the existence of a clopen subset G of V that separates the two closed sets in (8) in the sense that

$$\vartheta^{-1}(H_1^-) \subseteq G \quad \text{and} \quad \vartheta^{-1}(H_2^-) \cap G = \varnothing. \tag{9}$$

The element s belongs to exactly one of the sets G and $\sim G$. If s is in G, then G is in N_s, by the definition of N_s; so the set $\vartheta(G \cap U)^-$ belongs to M_s, by (7), and this set does not contain r_2. In more detail, if r_2 belonged to the closure of $\vartheta(G \cap U)$, then every open set containing r_2 would have a non-empty intersection with $\vartheta(G \cap U)$. In particular, the open set H_2 would have a non-empty intersection with $\vartheta(G \cap U)$. It follows that $\vartheta^{-1}(H_2)$ would have a non-empty intersection with $G \cap U$ and therefore also with G, in contradiction to the right-hand equation in (9). A similar argument applies if s is assumed to be in $\sim G$: the assumption implies that the set $\vartheta(\sim G \cap U)^-$

is in M_s, and this set does not contain r_1. For if it did contain r_1, then H_1 would have a non-empty intersection with $\vartheta(\sim G \cap U)^-$, and therefore $\vartheta^{-1}(H_1)$ would have a non-empty intersection with $\sim G$, in contradiction to the left-hand inclusion in (9).

Define the function δ on each point s in V by taking $\delta(s)$ to be the unique point that belongs to the intersection of M_s. This function is a well-defined mapping from V to W, by the observations of the preceding two paragraphs. To see that δ is an extension of ϑ, assume that s belongs to U. In this case, s belongs to the intersection $F \cap U$ for every set F in N_s, so $\vartheta(s)$ belongs to $\vartheta(F \cap U)$ and therefore also to $\vartheta(F \cap U)^-$, for every set F in N_s. It follows that $\vartheta(s)$ is the unique element in the intersection of M_s, by (7). Consequently, $\delta(s) = \vartheta(s)$, by the definition of δ.

It remains to prove that δ is continuous. To this end, consider an arbitrary open subset H of W. It must be shown that the inverse image of H under δ is open in V, and for this it suffices to prove that every element s in $\delta^{-1}(H)$, belongs to some clopen set F that is included in $\delta^{-1}(H)$. Fix a point s in $\delta^{-1}(H)$, and observe that $\delta(s)$ belongs to H. The definition of $\delta(s)$ specifies that this value is the unique element in the intersection $\bigcap M_s$, so this intersection must be included in H. A routine compactness argument shows that the intersection of some finite subset of M_s must already be included in H (see Exercise 24 on p. 279 of [10]). The intersection of any finite subset of M_s includes an element from M_s, by the argument used to prove that M_s has the finite intersection property. Consequently, there is a clopen set F in N_s such that

$$\vartheta(F \cap U)^- \subseteq H, \qquad (10)$$

by (7). The element s belongs to F, by the definition of N_s and the fact that F is in N_s. To see that F is included in $\delta^{-1}(H)$, consider an arbitrary element t in F. Since F is in N_t, by the definition of N_t, the set $\vartheta(F \cap U)^-$ must belong to M_t, by (7) (with t in place of s). Consequently, the intersection of the sets in M_t is included in $\vartheta(F \cap U)^-$ and therefore also in H, by (10). The value $\delta(t)$ is the unique point in the intersection of M_t, so $\delta(t)$ belongs to H, and therefore t belongs to $\delta^{-1}(H)$. This is true for every element t in F, so F is included in $\delta^{-1}(H)$. $\square$

In view of Theorem 2.59, we know that the maximum compactification $\mathfrak{V}$ of a union space $\mathfrak{U}$ has the topology of the classic Stone-Čech compactification, that is to say, every continuous function ϑ from $\mathfrak{U}$

into a compact Hausdorff space W can be extended to a continuous function from $\mathfrak{V}$ into W. It may happen that a compact Hausdorff space W is the universe of a relational structure $\mathfrak{W}$, and that a continuous function ϑ from $\mathfrak{U}$ to $\mathfrak{W}$ is also a homomorphism (though not necessarily a bounded homomorphism). What topological conditions must the relations in $\mathfrak{W}$ satisfy in order for the continuous extension of ϑ to also be a homomorphism? The answer is that the relations in $\mathfrak{W}$ must be closed subsets of the appropriate product space. Define a *closed space* to be a relational structure $\mathfrak{W}$ with a compact Hausdorff topology such that the relations in $\mathfrak{W}$ are closed in the product topology, that is to say, if R is a relation of rank n in $\mathfrak{W}$, then R is a closed subset of the product space W^n. Observe that every relational space is a closed space, by Theorem 2.15 and the remark preceding that theorem.

Theorem 2.60. *If $\mathfrak{V}$ is the Stone-Čech compactification of the union $\mathfrak{U}$ of a disjoint system of relational spaces, then every continuous homomorphism from $\mathfrak{U}$ into a closed space $\mathfrak{W}$ can be extended to a continuous homomorphism from $\mathfrak{V}$ into $\mathfrak{W}$.*

Proof. Suppose ϑ is a continuous homomorphism from $\mathfrak{U}$ into a closed space $\mathfrak{W}$. The topology on $\mathfrak{W}$ is, by definition, compact and Hausdorff, so there is a uniquely determined continuous function δ from $\mathfrak{V}$ into $\mathfrak{W}$ that extends ϑ, by Theorem 2.59. It must be demonstrated that δ preserves the fundamental relations of $\mathfrak{V}$. Focus on the case of a ternary relation R.

Assume u, v, and w are elements in $\mathfrak{V}$ such that

$$R(u, v, w) \tag{1}$$

holds in $\mathfrak{V}$, and write

$$r = \delta(u), \qquad s = \delta(v), \qquad t = \delta(w). \tag{2}$$

The goal is to prove that $R(r, s, t)$ holds in $\mathfrak{W}$. The strategy is to show that every open subset of the product space

$$W \times W \times W \tag{3}$$

that contains the triple (r, s, t) must have a non-empty intersection with R. Since the relation R is a closed subset of (3), by the assumption that $\mathfrak{W}$ is a closed space, it then follows that (r, s, t) belongs to R, so that $R(r, s, t)$ does hold. The sets of the form

$$H_1 \times H_2 \times H_3, \tag{4}$$

with H_k open in $\mathfrak{W}$ for $k = 1, 2, 3$, form a base for the product topology on (3), so it suffices to prove that every set of form (4) which contains the triple (r, s, t) has a non-empty intersection with R.

Consider open sets H_1, H_2, and H_3 in $\mathfrak{W}$ that contain the points r, s, and t respectively. The inverse images $\delta^{-1}(H_1)$ and $\delta^{-1}(H_2)$ are open subsets of $\mathfrak{V}$ that contain the points u and v respectively, by (2) and the continuity of δ. The clopen sets form a base for the topology of $\mathfrak{V}$, so there must be clopen sets F and G in $\mathfrak{V}$ such that

$$u \in F \qquad \text{and} \qquad v \in G, \tag{5}$$

and

$$F \subseteq \delta^{-1}(H_1) \qquad \text{and} \qquad G \subseteq \delta^{-1}(H_2). \tag{6}$$

The image set

$$R^*(F \times G) = \{z \in V : R(x, y, z) \text{ for some } x \in F \text{ and } y \in G\} \tag{7}$$

is clopen in $\mathfrak{V}$, because $\mathfrak{V}$ is a relational space (see Definition 2.2); and the point w belongs to this set, by (1), (5) and (7). Also, the inverse image $\delta^{-1}(H_3)$ is open and contains w, by (2), the continuity of δ, and the assumption that t is in H_3. Consequently, the intersection

$$R^*(F \times G) \cap \delta^{-1}(H_3) \tag{8}$$

is an open set in $\mathfrak{V}$ that contains w.

The universe of $\mathfrak{U}$ is a dense subset of $\mathfrak{V}$, because $\mathfrak{V}$ is a compactification of $\mathfrak{U}$ (see Definition 2.48), so the open set in (8) must intersect the universe of $\mathfrak{U}$ in some point $\bar{w}$. Since $\bar{w}$ belongs to the set in (7), there must be points $\bar{u}$ in F and $\bar{v}$ in G such that

$$R(\bar{u}, \bar{v}, \bar{w}) \tag{9}$$

holds in $\mathfrak{V}$. Now $\mathfrak{U}$ is an inner subspace of $\mathfrak{V}$ (because $\mathfrak{V}$ is a compactification of $\mathfrak{U}$) and consequently $\mathfrak{U}$ is algebraically an inner substructure of $\mathfrak{V}$, by Definition 2.48. Since $\bar{w}$ belongs to $\mathfrak{U}$, it follows that the points $\bar{u}$ and $\bar{v}$ must also belong to $\mathfrak{U}$, and (9) must hold in $\mathfrak{U}$, by the definition of an inner substructure. The continuous function ϑ is assumed to be a homomorphism from $\mathfrak{U}$ to $\mathfrak{W}$, so (9) implies that

$$R(\vartheta(\bar{u}), \vartheta(\bar{v}), \vartheta(\bar{w}))$$

holds in $\mathfrak{W}$. The function δ agrees with ϑ on $\mathfrak{U}$, so

$$\delta(\bar{u}) = \vartheta(\bar{u}), \qquad \delta(\bar{v}) = \vartheta(\bar{v}), \qquad \delta(\bar{w}) = \vartheta(\bar{w}),$$

and therefore

$$R(\delta(\bar{u}), \delta(\bar{v}), \delta(\bar{w})) \tag{10}$$

holds in $\mathfrak{W}$.

The points $\bar{u}$ and $\bar{v}$ belong to the sets F and G respectively, so they must also belong to the inverse images $\delta^{-1}(H_1)$ and $\delta^{-1}(H_2)$ respectively, by (6). Consequently, $\delta(\bar{u})$ is in H_1 and $\delta(\bar{v})$ in H_2, by the definition of the inverse image of a set. Also, the point $\bar{w}$ belongs to $\delta^{-1}(H_3)$, so $\delta(\bar{w})$ belongs to H_3. It follows that the triple

$$(\delta(\bar{u}), \delta(\bar{v}), \delta(\bar{w}))$$

belongs to the set in (4), and therefore to the intersection of this set with R, by (10). Thus, the set in (4) has a non-empty intersection with R, as was to be shown. Conclusion: the function δ preserves the fundamental relations of the relational structures, so it is a homomorphism; consequently, δ is a continuous homomorphism from $\mathfrak{V}$ to $\mathfrak{W}$ that extends ϑ. $\square$

Ian Hodkinson has kindly pointed out to us the following theorem due to Goldblatt that is apparently related to some of the results in this section: the ultrafilter space of the direct limit of a direct system of modal algebras is isomorphic to the inverse limit of the inverse system of ultrafilter spaces of the modal algebras in the direct system. According to Hodkinson, this theorem is a consequence of Theorems 10.7, 11.2, and 11.6 in [12].

Chapter 3
Hybrid Duality

A classic result in the theory of Boolean algebras states that the dual space of the Boolean algebra of all subsets of a set U is homeomorphic to the Stone-Čech compactification of the discrete space U, that is to say, U endowed with the discrete topology in which every subset is defined to be open (and hence closed). It is natural to try to extend this result to Boolean algebras with operators by showing that the dual relational space of the complex algebra $\mathfrak{Cm}(U)$ of an arbitrary relational structure $\mathfrak{U}$ is homeo-isomorphic to the Stone-Čech compactification of the discrete space $\mathfrak{U}$, that is to say, $\mathfrak{U}$ endowed with the discrete topology. Unfortunately, this extension fails to be true in general, as we shall eventually see, although it does hold for a special class of relational structures that are of finitary character (in a sense to be defined). Nevertheless, there is a characterization of the dual space of $\mathfrak{Cm}(U)$ that may be viewed as a generalization of the Boolean algebraic result: the dual space of $\mathfrak{Cm}(U)$ is homeo-isomorphic to a Stone-Čech weak compactification of the discrete space $\mathfrak{U}$. In fact, there is a lattice isomorphism between the lattice of equivalence classes of weak compactifications of the discrete space $\mathfrak{U}$ and the lattice of subalgebras of $\mathfrak{Cm}(U)$ that contain all singleton subsets; and this lattice isomorphism maps the Stone-Čech weak compactification of $\mathfrak{U}$ to the complex algebra $\mathfrak{Cm}(U)$ itself.

If a relational structure is of finitary character, then weak compactifications turn out to be the same thing as compactifications for the corresponding discrete space. In this case, the preceding results hold with the term "weak compactification" replaced by "compactification". In particular, for Boolean algebras (without operators) we obtain the apparently new result that there is an isomorphism from the lattice

S. Givant, *Duality Theories for Boolean Algebras with Operators*,
Springer Monographs in Mathematics, DOI 10.1007/978-3-319-06743-8_3,
© Springer International Publishing Switzerland 2014

of equivalence classes of compactifications of a discrete space U to the lattice of subalgebras of $Sb(U)$—the Boolean algebra of all subsets of U—that contain the finite subsets of U.

The arguments needed to establish these results are based upon a hybrid duality theory for homomorphisms that combines aspects of the algebraic and topological duality theories developed in Chapters 1 and 2 respectively. The notion of a bounded homomorphism is replaced by a weaker notion, namely that of a weakly bounded homomorphism from a relational *structure* $\mathfrak{U}$ into a relational *space* $\mathfrak{V}$. We show that every weakly bounded homomorphism from $\mathfrak{U}$ into $\mathfrak{V}$ has a dual that is a homomorphism from the (topologically defined) dual Boolean algebra with operators of $\mathfrak{V}$ to the (algebraically defined) dual complex algebra $\mathfrak{Cm}(U)$ of $\mathfrak{U}$, and vice versa; and each of these morphisms is its own second dual. The epi-mono duality no longer holds; duals of epimorphisms are monomorphisms, but duals of monomorphisms may fail to be epimorphisms.

3.1 Duality for Homomorphisms

The notion of a bounded homomorphism between relational structures (see Definition 1.8) is too strong for the purposes of this chapter, so a weaker version of the definition is introduced that is applicable when the range structure is not just a relational structure, but in fact a relational space, that is to say, a relational structure endowed with the topology of a Boolean space under which the fundamental relations are continuous and clopen (see Definition 2.2).

Definition 3.1. A homomorphism ϑ from a relational structure $\mathfrak{U}$ into a relational space $\mathfrak{V}$ is said to be *weakly bounded* if for every fundamental relation R of rank $n+1$, and every sequence of elements $r_0, \ldots, r_{n-1}$ in $\mathfrak{V}$ and element w in $\mathfrak{U}$, if $R(r_0, \ldots, r_{n-1}, \vartheta(w))$ holds in $\mathfrak{V}$, then for every sequence of clopen subsets $F_0, \ldots, F_{n-1}$ of $\mathfrak{V}$ such that F_i contains r_i for each i, there is an element u_i in $\vartheta^{-1}(F_i)$ for each i such that $R(u_0, \ldots, u_{n-1}, w)$ holds in $\mathfrak{U}$. $\square$

In the case of a ternary relation R, the preceding definition requires that for all elements r and s in $\mathfrak{V}$ and w in $\mathfrak{U}$, if $R(r, s, \vartheta(w))$ holds in $\mathfrak{V}$, then for every pair of clopen subsets F and G of $\mathfrak{V}$ containing r and s respectively, there must be elements u in $\vartheta^{-1}(F)$ and v in $\vartheta^{-1}(G)$ such that $R(u, v, w)$ holds in $\mathfrak{U}$.

The difference between the definition of a bounded homomorphism and that of a weakly bounded homomorphism is that in the former we require the existence of a sequence $u_0, \ldots, u_{n-1}$ of elements in $\mathfrak{U}$ satisfying the strong condition

$$\vartheta(u_i) = r_i$$

for each i (so the clopen sets $F_0, \ldots, F_{n-1}$ don't really play a role), whereas in the latter we do not require that $\vartheta(u_i)$ satisfy this strong condition, but rather the weaker condition that $\vartheta(u_i)$ belong to F_i for each i whenever $F_0, \ldots, F_{n-1}$ is an appropriate sequence of clopen sets; in particular, the elements u_i may vary as the clopen sets F_i vary (for fixed elements r_i). It is obvious that a bounded homomorphism is always weakly bounded, but the converse is in general not true, as we shall see after Theorem 3.4.

Goldblatt has pointed out that the notion of a weakly bounded homomorphism is closely related to a notion introduced by him in [14], namely the notion of a function on modal frames that has the strong forth property and the back property. Modal frames are relational structures with a single binary relation, together with a Boolean algebra of subsets (of the universe of the structure) that is closed under the formation of images of sets under the binary relation of the structure.

The next theorem is the hybrid analogue of Theorems 1.9 and 2.16.

Theorem 3.2. *Let $\mathfrak{U}$ be a relational structure, and $\mathfrak{V}$ a relational space with dual algebra $\mathfrak{B}$. If ϑ is a weakly bounded homomorphism from $\mathfrak{U}$ into $\mathfrak{V}$, then the function φ defined on elements F in $\mathfrak{B}$ by*

$$\varphi(F) = \vartheta^{-1}(F) = \{u \in U : \vartheta(u) \in F\}$$

is a homomorphism from $\mathfrak{B}$ into $\mathfrak{Cm}(U)$.

Proof. The proof that the function φ defined in the statement of the theorem is a Boolean homomorphism is identical to the proof of the corresponding part of Theorem 1.9, and is left to the reader. To see that φ preserves the operators of the algebras, consider the case of a binary operator $\circ$ that is defined in terms of a ternary relation R. Let F and G be elements in $\mathfrak{B}$, that is to say, clopen subsets of $\mathfrak{V}$. It must be shown that

$$\varphi(F \circ G) = \varphi(F) \circ \varphi(G), \tag{1}$$

or what amounts to the same thing, that

$$\vartheta^{-1}(F \circ G) = \vartheta^{-1}(F) \circ \vartheta^{-1}(G). \tag{2}$$

Assume first that an element w belongs to the left side of (2). In this case, $\vartheta(w)$ must be in $F \circ G$, by the definition of the inverse image of a set under a function, so there must be elements r in F and s in G such that

$$R(r, s, \vartheta(w)) \tag{3}$$

holds in $\mathfrak{V}$, by the definition of the operator $\circ$ in $\mathfrak{B}$ and in $\mathfrak{Cm}(V)$ (the former being a subalgebra of the latter). Since ϑ is assumed to be weakly bounded, there must be elements u in $\vartheta^{-1}(F)$ and v in $\vartheta^{-1}(G)$ such that

$$R(u, v, w) \tag{4}$$

holds in $\mathfrak{U}$. It follows from (4) and the definition of the operator $\circ$ in $\mathfrak{Cm}(U)$ that w belongs to the right side of (2).

Assume now that w belongs to the right side of (2). In this case there are elements u in $\vartheta^{-1}(F)$ and v in $\vartheta^{-1}(G)$ such that (4) holds in $\mathfrak{U}$, by the definition of the operator $\circ$ in $\mathfrak{Cm}(U)$. The homomorphism properties of ϑ imply that

$$R(\vartheta(u), \vartheta(v), \vartheta(w))$$

holds in $\mathfrak{V}$. Write $r = \vartheta(u)$ and $s = \vartheta(v)$ to arrive at the conclusion that r is in F, and s in G, and (3) holds in $\mathfrak{V}$. It follows that $\vartheta(w)$ is in the set $F \circ G$, by the definition of the operator $\circ$ in $\mathfrak{B}$, so w belongs to the left side of (2). This completes the proof of (2) and hence also of (1). Conclusion: φ is a homomorphism from $\mathfrak{B}$ into $\mathfrak{Cm}(U)$. $\square$

One difference between Theorems 3.2 and 1.9 is that in the latter, we obtain a complete homomorphism from $\mathfrak{Cm}(V)$ into $\mathfrak{Cm}(U)$, whereas in the former we obtain only a homomorphism from the subalgebra $\mathfrak{B}$ of clopen sets in $\mathfrak{Cm}(V)$ into $\mathfrak{Cm}(U)$. The corresponding difference between Theorems 3.2 and 2.16 is that in the latter an assumption of continuity is needed on the mapping ϑ, whereas in the former no such assumption is needed. Another difference between Theorem 3.2 and the earlier theorems is that Theorem 3.2 does not mention the epi/mono duality. In fact, if ϑ is one-to-one, it does not follow that φ is necessarily onto, and if φ is one-to-one, it does not follow that ϑ is necessarily onto. We shall discuss this point more fully after Theorem 3.4 below.

There is a kind of converse to Theorem 3.2 that is the analogue of Theorem 1.10.

Theorem 3.3. *Let $\mathfrak{U}$ be a relational structure, and $\mathfrak{V}$ a relational space with dual algebra $\mathfrak{B}$. If φ is a homomorphism from $\mathfrak{B}$ into $\mathfrak{Cm}(U)$, then the function ϑ defined on elements in $\mathfrak{U}$ by*

$$\vartheta(u) = r \quad \text{if and only if} \quad r \in \bigcap\{F \in B : u \in \varphi(F)\}$$

is a weakly bounded homomorphism from $\mathfrak{U}$ into $\mathfrak{V}$.

Proof. Fix an element u in $\mathfrak{U}$, and write

$$X_u = \{F \in B : u \in \varphi(F)\}. \tag{1}$$

Using this notation, the definition of ϑ assumes the form

$$\vartheta(u) = r \quad \text{if and only if} \quad r \in \bigcap X_u. \tag{2}$$

To see that ϑ is well defined, observe that X_u is an ultrafilter in $\mathfrak{B}$. (The argument is identical to the argument given in the corresponding part of the proof of Theorem 1.10, with F and X interchanged.) Consequently, there is a unique element r that belongs to the intersection of the sets in X_u. In more detail, the sets in X_u are clopen in $\mathfrak{V}$, by the definition of $\mathfrak{B}$. They have the finite intersection property, because X_u is an ultrafilter and is therefore closed under finite intersections and does not contain the empty set. It follows that the intersection $\bigcap X_u$ is non-empty, by the compactness of the space $\mathfrak{V}$. To see that no more than one element can belong to this intersection, assume r and s are points in $\mathfrak{V}$. Since $\mathfrak{V}$ is a Boolean space, there is a clopen set F that contains r but not s. If u is in $\varphi(F)$, then F belongs to X_u, and consequently s cannot be in $\bigcap X_u$. On the other hand, if u is not in $\varphi(F)$, then u must belong to the complement $\sim\varphi(F)$. Since

$$\sim\varphi(F) = \varphi(\sim F),$$

by the homomorphism properties of φ, it follows that $\sim F$ is in X_u, and therefore r cannot belong to $\bigcap X_u$. Thus, at most one of the two points r and s is in $\bigcap X_u$.

In order to formulate the definition of ϑ in a slightly different way, write

$$Y_r = \{F \in D : r \in F\} \tag{3}$$

and observe that Y_r is an ultrafilter in $\mathfrak{B}$. The condition in (2) that r belong to $\bigcap X_u$ is equivalent to the condition that every set in X_u

belong to Y_r. Since both X_u and Y_r are ultrafilters in $\mathfrak{B}$, this can only happen if $X_u = Y_r$. Thus,

$$\vartheta(u) = r \qquad \text{if and only if} \qquad X_u = Y_r. \tag{4}$$

Turn now to the task of showing that ϑ is a weakly bounded homomorphism. Consider as an example the case of a ternary relation R and a corresponding binary operator $\circ$. Let u, v, and w be elements in $\mathfrak{U}$, and write

$$\vartheta(u) = r, \qquad \vartheta(v) = s, \qquad \vartheta(w) = t.$$

From (4) it follows that

$$X_u = Y_r, \qquad X_v = Y_s, \qquad X_w = Y_t. \tag{5}$$

Assume $R(u, v, w)$ holds in $\mathfrak{U}$, with the goal of showing that $R(r, s, t)$ holds in $\mathfrak{V}$. The assumption implies that

$$w \in \{u\} \circ \{v\}, \tag{6}$$

by the definition of the operator $\circ$ in $\mathfrak{Cm}(U)$. From (1) it is clear that for every set F in X_u and every set G in X_v we have

$$u \in \varphi(F) \qquad \text{and} \qquad v \in \varphi(G),$$

or, put another way,

$$\{u\} \subseteq \varphi(F) \qquad \text{and} \qquad \{v\} \subseteq \varphi(G).$$

The monotony of the operator $\circ$ and the homomorphism properties of φ therefore imply that

$$\{u\} \circ \{v\} \subseteq \varphi(F) \circ \varphi(G) = \varphi(F \circ G). \tag{7}$$

Together, (6) and (7) show that w belongs to the set $\varphi(F \circ G)$, so $F \circ G$ is in X_w, by (1) (with w in place of u). This is true for every set F in X_u and every set G in X_v, so

$$X_u \circ X_v \subseteq X_w,$$

where the operation on the left side of this inclusion is the complex operation that is induced on subsets of $\mathfrak{B}$ by the binary operator $\circ$ in $\mathfrak{B}$. Combine this inclusion with (5) to obtain

$$Y_r \circ Y_s \subseteq Y_t. \tag{8}$$

The ultrafilters involved in this inclusion all belong to the dual space of $\mathfrak{B}$, which we denote by $\mathfrak{V}^*$. It follows from (8) and from the definition of the relation R in the dual space $\mathfrak{V}^*$ that $R(Y_r, Y_s, Y_t)$ holds in $\mathfrak{V}^*$. The relational space $\mathfrak{V}^*$ is the second dual of $\mathfrak{V}$, so the function δ mapping each point p in $\mathfrak{V}$ to the point Y_p in $\mathfrak{V}^*$ is a homeo-isomorphism from $\mathfrak{V}$ to $\mathfrak{V}^*$, by Theorem 2.11. It follows that $R(r, s, t)$ must hold in $\mathfrak{V}$, so ϑ is a homomorphism.

It remains to check that ϑ is weakly bounded as a homomorphism. To this end, consider elements r and s in $\mathfrak{V}$ and w in $\mathfrak{U}$. Write $\vartheta(w) = t$, and observe that

$$X_w = Y_t, \tag{9}$$

by (4). Assume $R(r, s, \vartheta(w))$ holds in $\mathfrak{V}$, that is to say, assume $R(r, s, t)$ holds in $\mathfrak{V}$. In this case, $R(Y_r, Y_s, Y_t)$ holds in $\mathfrak{V}^*$, by the isomorphism properties of the mapping δ, so the inclusion in (8) is valid, by the definition of the relation R in $\mathfrak{V}^*$. Let F and G be any pair of clopen sets in $\mathfrak{V}$ that contain the elements r and s respectively. The set F is in Y_r, and G is in Y_s, by (3), so the product $F \circ G$ must be in Y_t, by (8), and therefore also in X_w, by (9). It follows from this observation and (1) (with w in place of u) that w must belong to $\varphi(F \circ G)$, and therefore also to $\varphi(F) \circ \varphi(G)$, by the homomorphism properties of φ. Consequently, there must be elements u in $\varphi(F)$ and v in $\varphi(G)$ such that $R(u, v, w)$ holds in $\mathfrak{U}$, by the definition of the operator $\circ$ in $\mathfrak{Cm}(U)$. Since u is in $\varphi(F)$, the set F must belong to ultrafilter X_u, by (1). The element $\vartheta(u)$ belongs to every set in X_u, by the very definition of ϑ. In particular, $\vartheta(u)$ is in F and therefore u is in $\vartheta^{-1}(F)$. An analogous argument shows that v is in $\vartheta^{-1}(G)$. Thus, the homomorphism ϑ is weakly bounded, by Definition 3.1. $\square$

The homomorphism φ in the statement of Theorem 3.2 is called the (*first*) *dual*, or the *dual homomorphism*, of the given weakly bounded homomorphism ϑ. Notice that the definition of φ may be written in the form

$$u \in \varphi(F) \qquad \text{if and only if} \qquad \vartheta(u) \in F$$

for every element u in $\mathfrak{U}$ and every set F in $\mathfrak{B}$. The weakly bounded homomorphism ϑ in the statement of Theorem 3.3 is called the (*first*) *dual*, or the *dual weakly bounded homomorphism*, of the given homomorphism φ.

We continue with the assumption that $\mathfrak{U}$ is a relational structure, and $\mathfrak{V}$ a relational space with dual algebra $\mathfrak{B}$. Suppose φ is a homomorphism from $\mathfrak{B}$ into $\mathfrak{Cm}(U)$, and ϑ is the dual of φ. For a given element u in $\mathfrak{U}$, write $\vartheta(u) = r$, and observe that $X_u = Y_r$, by the equivalence in (4) in the proof of Theorem 3.3. Consequently,

$$u \in \varphi(F) \qquad \text{if and only if} \qquad F \in X_u,$$
$$\text{if and only if} \qquad F \in Y_r,$$
$$\text{if and only if} \qquad r \in F,$$
$$\text{if and only if} \qquad \vartheta(u) \in F,$$

for every element u in $\mathfrak{U}$ and every set F in $\mathfrak{B}$, by the definitions of the ultrafilters X_u and Y_r. Put another way,

$$u \in \varphi(F) \qquad \text{if and only if} \qquad u \in \vartheta^{-1}(F),$$

which is equivalent to saying that

$$\varphi(F) = \vartheta^{-1}(F)$$

for every set F in $\mathfrak{B}$. This equation is of course the definition of φ in terms of ϑ in Theorem 3.2. The point is that the equation is also valid in the context of Theorem 3.3, where φ is given and ϑ is being defined.

If φ is a homomorphism from $\mathfrak{B}$ into $\mathfrak{Cm}(U)$, and if ϑ is the dual weakly bounded homomorphism from $\mathfrak{U}$ into $\mathfrak{V}$, then the dual of ϑ is called the *second dual* of φ. It is the function φ^* from $\mathfrak{B}$ into $\mathfrak{Cm}(U)$ that is defined on sets F in $\mathfrak{B}$ by

$$\varphi^*(F) = \vartheta^{-1}(F),$$

by Theorem 3.2. Since the set $\vartheta^{-1}(F)$ coincides with $\varphi(F)$, by the observation at the end of the preceding paragraph, it follows that

$$\varphi^*(F) = \varphi(F)$$

for all sets F in $\mathfrak{B}$. Thus, φ is its own second dual.

Start now with a weakly bounded homomorphism ϑ from $\mathfrak{U}$ into $\mathfrak{V}$. If φ is the dual homomorphism from $\mathfrak{B}$ into $\mathfrak{Cm}(U)$, then the dual of φ is called the *second dual* of ϑ. It is the function ϑ^* whose value on an element u in $\mathfrak{U}$ is determined by the equivalence

$$\vartheta^*(u) = r \qquad \text{if and only if} \qquad X_u = Y_r,$$

by step (4) in the proof of Theorem 3.3. For every set F in $\mathfrak{B}$, we have

$$\varphi(F) = \vartheta^{-1}(F),$$

by the definition of φ as the dual of ϑ, so

$$
\begin{aligned}
F \in X_u \quad &\text{if and only if} \quad u \in \varphi(F), \\
&\text{if and only if} \quad u \in \vartheta^{-1}(F), \\
&\text{if and only if} \quad \vartheta(u) \in F, \\
&\text{if and only if} \quad F \in Y_{\vartheta(u)},
\end{aligned}
$$

by the definitions of the ultrafilters X_u and $Y_{\vartheta(u)}$. In other words, these two ultrafilters are equal. If $\vartheta^*(u) = r$, then

$$Y_r = X_u = Y_{\vartheta(u)},$$

and therefore $\vartheta^*(u) = r = \vartheta(u)$. This equality holds for every element u in $\mathfrak{U}$, so the function ϑ^* coincides with ϑ, and therefore ϑ is its own second dual.

The earlier dualities that we have studied carry with them a duality between one-to-one mappings and onto mappings. In the present context, that duality fails, as we shall see in a moment. In order to characterize the dual of a one-to-one mapping and the dual of an onto mapping, it is helpful to recall and generalize some notions that played a role in the topological discussion of Chapter 2. A subset W of a set V is said to be *dense* in V with respect to a collection Z of subsets of V if every non-empty set in Z contains an element from W. For example, in a Boolean space V, a subset W is dense in the standard topological sense if every non-empty clopen set (equivalently, if every non-empty open set) contains an element from W. In this example, Z is of course the collection of clopen subsets of V (equivalently, Z is the collection of open subsets of V). The set W is said to be *dense* in a collection Z of sets of subsets of V if for every set X in Z, there is an element in W that belongs to every set in X (or, equivalently, that belongs to the intersection of X). Thus, W is dense in the collection Z just in case W is dense in V with respect to the collection of sets

$$\{\bigcap X : X \in Z\}.$$

A set W of subsets of a set V is said to *separate points* in V if for any two distinct points in V, there is a set X in W that contains one of the two points, but not the other. For example, in a Boolean space V, the set W of clopen subsets of V separates points. The set W is said to *separate disjoint sets* in a collection Z of subsets of V if for any two disjoint sets in Z, there is a set in W that includes one of the two sets and is disjoint from the other. For example, in a Boolean space, the clopen sets separate disjoint closed subsets of the space. In this example, W is the collection of clopen subsets of V, and Z is the collection of closed subsets of V.

The next theorem summarizes the results obtained so far concerning hybrid duality and gives an appropriate substitute for the epi-mono duality.

Theorem 3.4. *Let $\mathfrak{U}$ be a relational structure, and $\mathfrak{V}$ a relational space with dual algebra $\mathfrak{B}$. There is a bijective correspondence between the set of weakly bounded homomorphisms ϑ from $\mathfrak{U}$ into $\mathfrak{V}$ and the set of homomorphisms φ from $\mathfrak{B}$ into $\mathfrak{Cm}(U)$ such that the equivalence*

$$u \in \varphi(F) \qquad \text{if and only if} \qquad \vartheta(u) \in F$$

holds for all sets F in $\mathfrak{B}$ and all elements u in $\mathfrak{U}$. Each of the mappings φ and ϑ is its own second dual. The function φ is one-to-one if and only if the image of U under the dual mapping ϑ, that is to say, the set

$$\vartheta(U) = \{\vartheta(u) : u \in U\},$$

is dense in the space $\mathfrak{V}$; and ϑ is one-to-one if and only if the set

$$\varphi(B) = \{\varphi(F) : F \in B\}$$

separates points in $\mathfrak{U}$. The mapping φ is onto if and only if the set of inverse images under ϑ of clopen subsets of $\mathfrak{V}$ separates disjoint subsets of $\mathfrak{U}$; and the mapping ϑ is onto if and only if U is dense in the collection

$$\{\varphi(Y_r) : r \in V\},$$

where

$$Y_r = \{F \in B : r \in F\} \qquad \text{and} \qquad \varphi(Y_r) = \{\varphi(F) : F \in Y_r\}.$$

Proof. The first two statements of the theorem were already proved in Theorems 3.2 and 3.3, and in the remarks following Theorem 3.3.

Turn therefore to the statements concerning the duals of one-to-one and of onto mappings. Fix a weakly bounded homomorphism ϑ from $\mathfrak{U}$ into $\mathfrak{V}$, and let φ be the dual homomorphism of ϑ that maps $\mathfrak{B}$ into $\mathfrak{Cm}(U)$. To prove that φ is one-to-one if and only if $\vartheta(U)$ is dense in $\mathfrak{V}$, we formulate six assertions, each of which is easily seen to be equivalent to its neighbor: (1) φ is one-to-one; (2) $\varphi(F) \neq \varnothing$ for every non-zero element F in $\mathfrak{B}$; (3) $\varphi(F) \neq \varnothing$ for every non-empty clopen subset F of $\mathfrak{V}$; (4) $\vartheta^{-1}(F) \neq \varnothing$ for every non-empty clopen subset F of $\mathfrak{V}$; (5) for every non-empty clopen subset F of $\mathfrak{V}$, there is an element u in $\mathfrak{U}$ such that $\vartheta(u)$ is in F; (6) $\vartheta(U)$ is dense in $\mathfrak{V}$. The equivalence of (1) and (2) follows from the fact that φ is a Boolean homomorphism, and a Boolean homomorphisms is one-to-one if and only if its kernel consists only of the empty set; the equivalence of (2) and (3) follows from the definition of $\mathfrak{B}$ as the dual algebra of $\mathfrak{V}$; the equivalence of (3) and (4) follows from the definition of φ as the dual homomorphism of ϑ; the equivalence of (4) and (5) follows from the definition of the inverse image of a set under a function; and the equivalence of (5) and (6) follows from the definition of density.

To prove that ϑ is one-to-one if and only if $\varphi(B)$ separates points in $\mathfrak{U}$, let u and v be distinct points in $\mathfrak{U}$, and consider the following four assertions, each of which is equivalent to its neighbor: (1) $\vartheta(u) \neq \vartheta(v)$; (2) there is a clopen subset F of $\mathfrak{V}$ that contains one of $\vartheta(u)$ and $\vartheta(v)$, but not the other; (3) there is a clopen subset F of $\mathfrak{V}$ such that $\vartheta^{-1}(F)$ contains one of u and v but not the other; (4) there is an element F in $\mathfrak{B}$ such that $\varphi(F)$ contains one of u and v, but not the other. The equivalence of (1) and (2) follows from the fact that in a Boolean space, the clopen sets separate points; the equivalence of (2) and (3) follows from the definition of the inverse image of a set under a function; and the equivalence of (3) and (4) follows from the definitions of φ and $\mathfrak{B}$ as the duals of ϑ and $\mathfrak{V}$ respectively. The function ϑ is one-to-one if and only if (1) holds for all distinct u and v in $\mathfrak{U}$, and the set $\varphi(B)$ separates points in $\mathfrak{U}$ if and only if (4) holds for all distinct u and v in $\mathfrak{U}$, so ϑ is one-to-one if and only if $\varphi(B)$ separates points.

To prove that φ is onto if and only if the inverse images under ϑ of clopen sets in $\mathfrak{V}$ separate disjoint subsets of $\mathfrak{U}$, consider the following three equivalent assertions about a subset X of $\mathfrak{U}$: (1) $X = \varphi(F)$ for some element F in $\mathfrak{B}$; (2) $X = \vartheta^{-1}(F)$ for some clopen subset F of $\mathfrak{V}$; (3) if Y is a subset of $\mathfrak{U}$ that is disjoint from X, then Y is disjoint from $\vartheta^{-1}(F)$, and X is included in $\vartheta^{-1}(F)$, for some clopen subset F of $\mathfrak{V}$. The equivalence of (1) and (2) follows from the definitions of the dual mapping φ and the dual algebra $\mathfrak{B}$, and the implication from (2) to (3) is obvious. To establish the implication from (3) to (2), assume that (3) holds, and take $Y = \sim X$. By (3), there must be a clopen subset F of $\mathfrak{V}$ such that the inverse image $\vartheta^{-1}(F)$ includes X and is disjoint from $Y = \sim X$. In other words,

$$X \subseteq \vartheta^{-1}(F) \qquad \text{and} \qquad \vartheta^{-1}(F) \subseteq X,$$

so $X = \vartheta^{-1}(F)$. Thus, (2) holds. The function φ is onto if and only if (1) holds for every subset X of $\mathfrak{U}$; and the inverse images under ϑ of clopen subsets of $\mathfrak{V}$ separate disjoint subsets of $\mathfrak{U}$ if and only if (3) holds for every subset X of $\mathfrak{U}$. Thus, φ is onto if and only if the inverse images in question separate disjoint subsets of $\mathfrak{U}$.

To prove that ϑ is onto if and only if the set U is dense in the collection

$$\Phi = \{\varphi(Y_r) : r \in V\},$$

fix elements u in $\mathfrak{U}$ and r in $\mathfrak{V}$, and consider the following four equivalent statements: (1) $\vartheta(u) = r$; (2) $\vartheta(u)$ is in F for every set F in Y_r; (3) u is in $\vartheta^{-1}(F)$ for every F in Y_r; (4) u is in $\varphi(F)$ for every F in Y_r. Statement (1) implies (2) because of the definition of Y_r as the set of clopen sets that contain r. For the reverse implication, observe that r is the unique element that belongs to every set in Y_r (by the compactness of the space $\mathfrak{V}$ and the fact that clopen sets separate points in $\mathfrak{V}$). The equivalence of (2) and (3) follows from the definition of the inverse image of a set under a function, and the equivalence of (3) and (4) follows from the definition of φ. The function ϑ is onto if and only if for every element r in $\mathfrak{V}$ there is an element u in $\mathfrak{U}$ such that (1) holds; and the set U is dense in the collection Φ if and only if for every element r in V there is an element u in U such that (4) holds. Consequently, ϑ is onto just in case U is dense in Φ.

The preceding characterizations of one-to-oneness and ontoness have been established under the hypothesis that ϑ is an arbitrary weakly bounded homomorphism from $\mathfrak{U}$ into $\mathfrak{V}$, and φ is the homomorphism from $\mathfrak{B}$ into $\mathfrak{Cm}(U)$ that is the dual of ϑ. Assume now that φ is an arbitrary homomorphism from $\mathfrak{B}$ into $\mathfrak{Cm}(U)$, and let ϑ be the weakly bounded homomorphism from $\mathfrak{U}$ into $\mathfrak{V}$ that is the dual of φ. The dual of ϑ is the second dual of φ, which is just φ, by the second statement of the theorem. The characterizations that were established in the preceding paragraphs may therefore be applied in the present case to obtain the desired results. $\square$

The full epi-mono dualities of Theorems 1.11 and 2.20 fail in the present case. For example, let $\mathfrak{V}$ be a relational space in which the topology is not discrete, that is to say, in which not every subset of $\mathfrak{V}$ is open. Take $\mathfrak{U}$ to be the relational structure that is the algebraic part of $\mathfrak{V}$. We construct a weakly bounded monomorphism ϑ from $\mathfrak{U}$ to $\mathfrak{V}$ with the property that the dual of ϑ is not onto. In fact, just take ϑ to be the identity function on $\mathfrak{U}$. Clearly, ϑ is a monomorphism— and actually an (algebraic) isomorphism—from $\mathfrak{U}$ to $\mathfrak{V}$. It is not difficult to check that ϑ is weakly bounded; in fact, ϑ is bounded, by Corollary 1.14, so ϑ must be weakly bounded. The dual of ϑ is the homomorphism φ from the dual algebra $\mathfrak{B}$ (of $\mathfrak{V}$) into $\mathfrak{Cm}(U)$ that is defined by $\varphi(F) = \vartheta^{-1}(F)$ for every F in $\mathfrak{B}$. We shall show that φ cannot be onto. From the definition of the inverse image of a set under a function, the assumption that ϑ is the identity function, and the assumption that $U = V$, it follows that

$$\vartheta^{-1}(F) = \{u \in U : \vartheta(u) \in F\} = \{u \in U : u \in F\} = U \cap F = F.$$

Consequently, φ is the identity function on the universe of $\mathfrak{B}$. The relational structure $\mathfrak{U}$ coincides with the algebraic part of $\mathfrak{V}$, so every set in $\mathfrak{B}$ must belong to $\mathfrak{Cm}(U)$. There are, however, sets in $\mathfrak{Cm}(U)$ that do not belong to $\mathfrak{B}$. This is because the topology on $\mathfrak{V}$ is assumed not to be discrete, and therefore there are subsets of U that are not clopen in the topology on $\mathfrak{V}$. It follows that φ maps $\mathfrak{B}$ onto a proper subalgebra of $\mathfrak{Cm}(U)$, namely $\mathfrak{B}$, and not onto $\mathfrak{Cm}(U)$.

In the next example, a monomorphism φ is constructed for which the dual weakly bounded homomorphism ϑ is not onto. The strategy of the construction is indirect: we create a weakly bounded homomorphism (and in fact a monomorphism) ϑ that is not onto, but whose dual φ is one-to-one. Let V be the set of all functions from the set $\mathbb{N}$ of natural numbers into $\{0, 1\}$. To describe a topology on V, write P for the set of (finite) functions whose domain is a finite set of natural numbers and whose range is included in $\{0, 1\}$. For each finite function p in P, say with domain D, the set of extensions of p that belong to V, that is to say, the set

$$F_p = \{r \in V : \text{ the restriction of } r \text{ to } D \text{ is } p\},$$

is declared to be a *basic clopen set*. The clopen subsets of V are defined to be the unions of finite systems of basic clopen sets, and the open subsets of V are defined to be the unions of arbitrary systems of basic clopen sets. It is not difficult to check that the collection of open sets so defined really is a topology on V, and that under this topology, V becomes a Boolean space (see Theorem 30 on p. 303 of [10]). Let R be the universal ternary relation on V, that is to say,

$$R = V \times V \times V.$$

Straightforward arguments show that R is clopen and continuous in the topology on V, so the relational structure

$$\mathfrak{V} = (V, R)$$

endowed with the topology described above is a relational space. In more detail, if F and G are subsets of V, then the image set

$$R^*(F \times G) = \{t \in V : R(r, s, t) \text{ for some } r \in F \text{ and } s \in G\}$$

is either V or empty, according to whether F and G are both non-empty, or at least one of them is empty. In either case, the image set is clopen (even when F and G are not clopen), so R is a clopen relation. Similarly, if H is a subset of V then the inverse image set

$$R^{-1}(H) = \{(r, s) \in V \times V : R(r, s, t) \text{ implies } t \in H\}$$

is either $V \times V$ or empty, according to whether H is, or is not, equal to V. In either case, the inverse image set is open in the product topology on $V \times V$ (even when H is not open), so R is a continuous relation.

A function r in V is said to have *finite support* if the set

$$\{n : n \in \mathbb{N} \text{ and } r(n) = 1\}$$

is finite. Take U to be the set of functions in V with finite support, and take $\mathfrak{U}$ to be the relational structure that is the restriction of $\mathfrak{V}$ to the set U. Thus, the fundamental relation in $\mathfrak{U}$ is just the universal ternary relation on the set U. A straightforward argument shows that U is a dense subset of $\mathfrak{V}$. Indeed, in view of the definition of the topology on V, it suffices to check that every basic clopen set has a non-empty intersection with U. Consider a finite function p in P, say with domain D, and define a function r in V by

$$r(n) = \begin{cases} p(n) & \text{if } n \in D, \\ 0 & \text{if } n \in \mathbb{N} \sim D. \end{cases}$$

Clearly, r is an extension of p that has finite support, so r belongs to the intersection of F_p and U. In particular, this intersection is not empty, so U is dense in V. Notice also that U is different from V. For instance, for any infinite set D of natural numbers, the function r defined by

$$r(n) = \begin{cases} 1 & \text{if} \quad n \in D, \\ 0 & \text{if} \quad n \in \mathbb{N} \sim D, \end{cases}$$

belongs to V but not to U.

We are now ready to define the weakly bounded homomorphism ϑ: it is simply the identity mapping on the universe of $\mathfrak{U}$. Obviously, ϑ is a monomorphism that maps $\mathfrak{U}$ to a proper substructure of the algebraic part of $\mathfrak{V}$ (namely, $\mathfrak{U}$), so ϑ is not onto. The monomorphism ϑ is not bounded, but it is weakly bounded. To see that ϑ is not bounded, consider any elements r and s that are in $\mathfrak{V}$ but not in $\mathfrak{U}$, and let w be any element in $\mathfrak{U}$. Clearly, $R(r, s, \vartheta(w))$ holds in $\mathfrak{V}$, since R is the universal ternary relation. However, there can be no elements u and v in $\mathfrak{U}$ such that $\vartheta(u) = r$ and $\vartheta(v) = s$, because ϑ is the identity function on $\mathfrak{U}$, and r and s do not belong to $\mathfrak{U}$. To see that ϑ is weakly bounded, suppose r and s are elements in $\mathfrak{V}$, and w an element in $\mathfrak{U}$, such that $R(r, s, \vartheta(w))$ holds in $\mathfrak{V}$. For arbitrary clopen subsets F and G of $\mathfrak{V}$ containing the elements r and s respectively, the inverse images

$$\vartheta^{-1}(F) = F \cap U \qquad \text{and} \qquad \vartheta^{-1}(G) = G \cap U$$

must be non-empty, by the density of the set U in $\mathfrak{V}$. Consequently, there are elements u and v in U belonging to the sets F and G respectively. Obviously, $R(u, v, w)$ holds in $\mathfrak{U}$ (and in $\mathfrak{V}$), by the definition of the fundamental relation in $\mathfrak{U}$ (and $\mathfrak{V}$). Therefore, ϑ is weakly bounded with respect to the relation R.

Let $\mathfrak{B}$ be the dual algebra of $\mathfrak{V}$. The dual of the weakly bounded homomorphism ϑ is the homomorphism φ from $\mathfrak{B}$ into $\mathfrak{Cm}(U)$ defined by

$$\varphi(F) = \vartheta^{-1}(F) = F \cap U$$

for sets F in $\mathfrak{B}$. This homomorphism is one-to-one, by Theorem 3.4, because the set $\vartheta(U)$—which is just U—is dense in $\mathfrak{V}$; but the dual of φ, which is the second dual of ϑ and is therefore just ϑ, is not onto.

Despite the negative results discussed in the preceding paragraphs regarding an epi-mono duality, we do have the following positive partial result.

Corollary 3.5. *For the dual functions ϑ and φ in Theorem 3.4, if ϑ is onto, then φ is one-to-one, and if φ is onto, then ϑ is one-to-one.*

Proof. Suppose first that the weakly bounded homomorphism ϑ is onto. The set U must then be dense in the collection

$$\{\varphi(Y_r) : r \in V\}, \tag{1}$$

by Theorem 3.4. If G is an arbitrary non-zero element in $\mathfrak{B}$, then there must be a point r in $\mathfrak{V}$ that belongs to G. Consequently, G belongs to Y_r, by the definition of Y_r. The density of U in the collection of sets in (1) implies the existence of an element u in U that belongs to every set in

$$\varphi(Y_r) = \{\varphi(F) : F \in Y_r\}.$$

In particular, u belongs to $\varphi(G)$, so $\varphi(G)$ is not empty. This argument shows that the image under φ of every non-zero element in $\mathfrak{B}$ is a non-zero element in $\mathfrak{Cm}(U)$, so φ is one-to-one.

Assume next that the homomorphism φ is onto. The set of inverse images

$$\{\vartheta^{-1}(F) : F \in B\} \tag{2}$$

must then separate disjoint subsets of $\mathfrak{U}$, by Theorem 3.4. If u and v are distinct elements in $\mathfrak{U}$, then the singletons of these two elements are disjoint subsets of $\mathfrak{U}$. Consequently, there must be a set in (2) that includes one of the two singletons and is disjoint from the other, say $\vartheta^{-1}(F)$ includes $\{u\}$ and is disjoint from $\{v\}$. It follows that $\vartheta(u)$ belongs to F, but $\vartheta(v)$ does not, so $\vartheta(u)$ and $\vartheta(v)$ must be distinct. Therefore, ϑ is one-to-one. $\square$

3.2 Duality for Subalgebras

There is a rather unusual duality between a class of subalgebras of the complex algebra of a relational structure $\mathfrak{U}$ and a class of relational spaces that are weak compactifications of $\mathfrak{U}$ endowed with the discrete topology. Somewhat surprisingly, the construction of this duality parallels the one given in Section 2.11, concerning subalgebras of a direct product and compactifications of union spaces. The first step is to introduce the notion of a weak inner substructure of a relational space and to clarify the relationship of this notion to that of an inner substructure (see Definition 1.12).

Definition 3.6. A relational structure $\mathfrak{U}$ is defined to be a *weak inner substructure* of a relational space $\mathfrak{V}$ if algebraically $\mathfrak{U}$ is a substructure of $\mathfrak{V}$ and if for every fundamental relation R of rank $n + 1 > 1$ in $\mathfrak{V}$, and every sequence of elements $r_0, \ldots, r_{n-1}$ in $\mathfrak{V}$ and t in $\mathfrak{U}$, if $R(r_0, \ldots, r_{n-1}, t)$ holds in $\mathfrak{V}$, then for every sequence of clopen subsets $F_0, \ldots, F_{n-1}$ of $\mathfrak{V}$ such that F_i contains r_i for each i, there is an element $\bar{r}_i$ in $F_i \cap U$ for each i such that $R(\bar{r}_0, \ldots, \bar{r}_{n-1}, t)$ holds in $\mathfrak{U}$ (and therefore also in $\mathfrak{V}$). □

In the case of a ternary relation R, the preceding definition requires that for all elements r and s in $\mathfrak{V}$ and t in $\mathfrak{U}$, if $R(r, s, t)$ holds in $\mathfrak{V}$, then for every pair of clopen subsets F and G of $\mathfrak{V}$ containing r and s respectively, there must be elements $\bar{r}$ in $F \cap U$ and $\bar{s}$ in $G \cap U$ such that $R(\bar{r}, \bar{s}, t)$ holds in $\mathfrak{U}$ (and therefore also in $\mathfrak{V}$).

The difference between the definition of an inner substructure and that of a weak inner substructure is that under the given hypothesis— namely that $R(r_0, \ldots, r_{n-1}, t)$ holds in $\mathfrak{V}$ and t is in $\mathfrak{U}$—in the former the elements r_i are themselves required to belong to $\mathfrak{U}$, whereas in the latter it is only required that each sequence of clopen subsets of $\mathfrak{V}$ containing the elements r_i must also contain "local approximations" to these elements in $\mathfrak{U}$, that is to say, for each index i, there must be an element $\bar{r}_i$ in the ith clopen set of the sequence that belongs to $\mathfrak{U}$ and such that $R(\bar{r}_0, \ldots, \bar{r}_{n-1}, t)$ holds in $\mathfrak{U}$.

Every inner substructure $\mathfrak{U}$ of a relational space $\mathfrak{V}$ is a weak inner substructure of $\mathfrak{V}$, but the converse fails in general. The converse does hold in the special case when $\mathfrak{U}$ is of *finitary character* in the sense that the set

$$\{(r_0, \ldots, r_{n-1}) \in U^n : R(r_0, \ldots, r_{n-1}, t)\}$$

is finite for every relation R in $\mathfrak{U}$ of rank $n + 1 > 1$ and every element t in $\mathfrak{U}$.

Lemma 3.7. *An inner substructure of a relational space $\mathfrak{V}$ is always a weak inner substructure of $\mathfrak{V}$. For relational structures of finitary character, the converse is also true: a weak inner substructure of $\mathfrak{V}$ of finitary character is necessarily an inner substructure of $\mathfrak{V}$.*

Proof. Focus on the case of a ternary relation R. Assume first that $\mathfrak{U}$ is an inner substructure of $\mathfrak{V}$. Let r and s be elements in $\mathfrak{V}$, and t an element in $\mathfrak{U}$, such that $R(r, s, t)$ holds in $\mathfrak{V}$. The elements r and s then belong to $\mathfrak{U}$, because $\mathfrak{U}$ is an inner substructure of $\mathfrak{V}$. Consequently, if F and G are any clopen sets in $\mathfrak{V}$ that contain r and s respectively, then there are certainly elements $\bar{r}$ in $F \cap U$ and $\bar{s}$ in $G \cap U$ such that $R(\bar{r}, \bar{s}, t)$ holds in $\mathfrak{U}$, namely the elements $\bar{r} = r$ and $\bar{s} = s$. Thus, $\mathfrak{U}$ is a weak inner substructure of $\mathfrak{V}$.

Assume now that $\mathfrak{U}$ is a weak inner substructure of $\mathfrak{V}$ of finitary character. Let r and s be elements in $\mathfrak{V}$, and t an element in $\mathfrak{U}$, such that $R(r, s, t)$ holds in $\mathfrak{V}$. The assumed finitary character of $\mathfrak{U}$ implies that the sets

$$X = \{\bar{r} \in U : R(\bar{r}, \bar{s}, t) \text{ for some } \bar{s} \in U\} \tag{1}$$

and

$$Y = \{\bar{s} \in U : R(\bar{r}, \bar{s}, t) \text{ for some } \bar{r} \in U\} \tag{2}$$

are finite. For each element $\bar{r}$ in X that is different from r, there must be a clopen subset $F_{\bar{r}}$ of $\mathfrak{V}$ that contains r but not $\bar{r}$, because the topology on $\mathfrak{V}$ is that of a Boolean space, and consequently the clopen sets separate points. The set

$$F = \bigcap\{F_{\bar{r}} : \bar{r} \in X \text{ and } \bar{r} \neq r\}$$

is the intersection of finitely many clopen sets containing r, so F must also be a clopen set that contains r. However, F cannot contain any points from X that are different from r, because each such point is excluded by one of the sets in the intersection. Similarly, for each element $\bar{s}$ in Y that is different from s, there is a clopen subset $G_{\bar{s}}$ of $\mathfrak{V}$ that contains s but not $\bar{s}$. The intersection

$$G = \bigcap\{G_{\bar{s}} : \bar{s} \in Y \text{ and } \bar{s} \neq s\}$$

is a clopen subset of $\mathfrak{V}$ that contains s, but no other points from Y. Now $R(r, s, t)$ holds in $\mathfrak{V}$, by assumption, so there must be elements $\bar{r}$ in $F \cap U$ and $\bar{s}$ in $G \cap U$ such that

$$R(\bar{r}, \bar{s}, t) \tag{3}$$

holds in $\mathfrak{U}$, because $\mathfrak{U}$ is assumed to be a weak inner substructure of $\mathfrak{V}$. It follows from (1)–(3) that $\bar{r}$ belongs to X and $\bar{s}$ to Y. Since $\bar{r}$ is in F and also in X, the definition of F implies that $\bar{r}$ must be equal to r. Similarly, since $\bar{s}$ is in G and also in Y, the definition of G implies that $\bar{s}$ must be equal to s. But $\bar{r}$ and $\bar{s}$ belong to $\mathfrak{U}$, so r and s must also belong $\mathfrak{U}$. Conclusion: $\mathfrak{U}$ is an inner substructure of $\mathfrak{V}$. $\quad\square$

Definition 3.6 is so constructed that weak inner substructures are just the images of relational structures under weakly bounded monomorphisms.

Lemma 3.8. *If ϑ is a weakly bounded homomorphism from a relational structure $\mathfrak{U}$ into a relational space $\mathfrak{V}$, then the restriction of $\mathfrak{V}$ to the range of ϑ is a weak inner substructure of $\mathfrak{V}$. In the reverse direction, if ϑ is an isomorphism from $\mathfrak{U}$ to a weak inner substructure of $\mathfrak{V}$, then ϑ is a weakly bounded monomorphism from $\mathfrak{U}$ into $\mathfrak{V}$.*

Proof. Focus on the case of a ternary relation R. Assume first that ϑ is a weakly bounded homomorphism from $\mathfrak{U}$ into $\mathfrak{V}$, and let $\mathfrak{W}$ be the restriction of $\mathfrak{V}$ to the range of ϑ. Clearly, $\mathfrak{W}$ is a substructure of $\mathfrak{V}$, by the definition of a restriction. To prove that $\mathfrak{W}$ is a weak inner substructure of $\mathfrak{V}$, consider elements r and s in $\mathfrak{V}$ and t in $\mathfrak{W}$, and let F and G be clopen sets in $\mathfrak{V}$ that contain r and s respectively. Since ϑ maps $\mathfrak{U}$ onto $\mathfrak{W}$, there must be an element w in $\mathfrak{U}$ such that $\vartheta(w) = t$. If $R(r, s, t)$ holds in $\mathfrak{V}$, then of course $R(r, s, \vartheta(w))$ holds, so there must be elements u in $\vartheta^{-1}(F)$ and v in $\vartheta^{-1}(G)$ such that $R(u, v, w)$ holds in $\mathfrak{U}$, by the assumption that ϑ is weakly bounded. Putting

$$\bar{r} = \vartheta(u) \qquad \text{and} \qquad \bar{s} = \vartheta(v), \tag{1}$$

we see that $\bar{r}$ is in $F \cap W$ and $\bar{s}$ in $G \cap W$, and $R(\bar{r}, \bar{s}, t)$ holds in $\mathfrak{W}$, by the homomorphism properties of ϑ. Thus, $\mathfrak{W}$ is a weak inner substructure of $\mathfrak{V}$, by Definition 3.6.

Assume now that $\mathfrak{W}$ is a weak inner substructure of $\mathfrak{V}$ and that ϑ is an isomorphism from $\mathfrak{U}$ to $\mathfrak{W}$. To show that ϑ is weakly bounded, consider elements r and s in $\mathfrak{V}$, and w in $\mathfrak{U}$, and let F and G be clopen sets in $\mathfrak{V}$ that contain r and s respectively. If $R(r, s, \vartheta(w))$ holds in $\mathfrak{V}$, then there must be elements $\bar{r}$ in $F \cap W$ and $\bar{s}$ in $G \cap W$ such that $R(\bar{r}, \bar{s}, \vartheta(w))$ holds in $\mathfrak{W}$, by the assumption that $\mathfrak{W}$ is a weak inner substructure of $\mathfrak{V}$. There exist elements u and v in $\mathfrak{U}$ such that (1) holds, and consequently such that $R(u, v, w)$ holds in $\mathfrak{U}$, by the isomorphism properties of ϑ. Clearly, u is in $\vartheta^{-1}(F)$ and v in $\vartheta^{-1}(G)$, so ϑ is weakly bounded. $\square$

By taking for ϑ the identity function on the universe of $\mathfrak{U}$ in the preceding lemma, we arrive at the following characterization of weak inner substructures.

Corollary 3.9. *A relational structure $\mathfrak{U}$ is a weak inner substructure of a relational space $\mathfrak{V}$ if and only if $\mathfrak{U}$ is a substructure of $\mathfrak{V}$ and the identity function on $\mathfrak{U}$ is weakly bounded as a monomorphism from $\mathfrak{U}$ into $\mathfrak{V}$.*

The next lemma is an analogue of Lemma 2.50. It says, in particular, that the (topological) dual algebras of relational spaces extending a relational structure $\mathfrak{U}$ can all be mapped homomorphically onto subalgebras of $\mathfrak{Cm}(U)$ via relativization.

Lemma 3.10. *If $\mathfrak{U}$ is weak inner substructure of a relational space $\mathfrak{V}$, and if $\mathfrak{B}$ is the dual algebra of $\mathfrak{V}$, then the set*

$$B_0 = \{F \cap U : F \in B\}$$

is a subuniverse of $\mathfrak{Cm}(U)$, and the function mapping F to $F \cap U$ for each F in $\mathfrak{B}$ is an epimorphism from $\mathfrak{B}$ to the corresponding subalgebra $\mathfrak{B}_0$. The epimorphism is an isomorphism if and only if U is dense in $\mathfrak{V}$.

Proof. The identity function on $\mathfrak{U}$—call it ϑ—is a weakly bounded monomorphism from $\mathfrak{U}$ into $\mathfrak{V}$, by Corollary 3.9 and the assumption that $\mathfrak{U}$ is a weak inner substructure of $\mathfrak{V}$. The dual of ϑ is the homomorphism ψ from $\mathfrak{B}$ into $\mathfrak{Cm}(U)$ that is defined by

$$\psi(F) = \vartheta^{-1}(F)$$

for elements F in $\mathfrak{B}$, by Theorem 3.2. Since

$$\vartheta^{-1}(F) = \{u \in U : \vartheta(u) \in F\} = \{u \in U : u \in F\} = F \cap U,$$

by the assumption that ϑ is the identity function on $\mathfrak{U}$, the definition of ψ may be written in the form

$$\psi(F) = F \cap U$$

for every F in $\mathfrak{B}$. It follows that the range of ψ is just the set B_0. The range of a homomorphism is obviously a subuniverse of the target algebra, so B_0 is a subuniverse of $\mathfrak{Cm}(U)$. If $\mathfrak{B}_0$ is the corresponding subalgebra, then ψ is an epimorphism from $\mathfrak{B}$ to $\mathfrak{B}_0$. The mapping ψ is one-to-one (and therefore an isomorphism) if and only if $\vartheta(U)$ is dense in $\mathfrak{V}$, by Theorem 3.4. Since ϑ is the identity function on U, it follows that ψ is an isomorphism if and only if U is dense in $\mathfrak{V}$.   □

We shall refer to the algebra $\mathfrak{B}_0$ in the preceding lemma as the *relativization* of $\mathfrak{B}$ to U, and we shall refer to the mapping ψ as the *relativization epimorphism* (or the *relativization isomorphism* if ψ is one-to-one). The lemma motivates the following definition.

Definition 3.11. A relational space $\mathfrak{V}$ is said to be a *topological extension* of a relational structure $\mathfrak{U}$ if $\mathfrak{U}$ is (algebraically) a weak inner substructure of $\mathfrak{V}$, and if the universe of $\mathfrak{U}$ is a dense subset of $\mathfrak{V}$. □

The most important special case of Lemma 3.10 can be reformulated in the following way: if $\mathfrak{B}$ is the dual algebra of a topological extension of a relational structure $\mathfrak{U}$, then $\mathfrak{B}$ is isomorphic via relativization to a subalgebra of $\mathfrak{Cm}(U)$. The next lemma characterizes the relation of inclusion between subalgebras of $\mathfrak{Cm}(U)$ in topological terms. It is an analogue of Lemma 2.51.

Lemma 3.12. *Let $\mathfrak{V}$ and $\mathfrak{W}$ be topological extensions of a relational structure $\mathfrak{U}$, and $\mathfrak{B}$ and $\mathfrak{C}$ their respective dual algebras. The relativization of $\mathfrak{C}$ to U is a subalgebra of the relativization of $\mathfrak{B}$ to U if and only if there is a continuous bounded epimorphism from $\mathfrak{V}$ to $\mathfrak{W}$ that is the identity function on $\mathfrak{U}$.*

Proof. Let $\mathfrak{B}_0$ be the relativization of $\mathfrak{B}$ to U, and ψ the corresponding relativization isomorphism defined by

$$\psi(F) = F \cap U \tag{1}$$

for each set F in $\mathfrak{B}$ (see Lemma 3.10). Similarly, let $\mathfrak{C}_0$ be the relativization of $\mathfrak{C}$ to U, and ϱ the corresponding relativization isomorphism defined by

$$\varrho(F) = F \cap U \tag{2}$$

for each set F in $\mathfrak{C}$.

Assume first that there is a continuous bounded epimorphism ϑ from $\mathfrak{V}$ to $\mathfrak{W}$ mapping each element in $\mathfrak{U}$ to itself, with the goal of showing that $\mathfrak{C}_0$ must be a subalgebra of $\mathfrak{B}_0$. (The proof is very similar to the corresponding part of the proof of Lemma 2.51.) The topological dual of ϑ is the monomorphism φ from $\mathfrak{C}$ to $\mathfrak{B}$ that is defined by

$$\varphi(F) = \vartheta^{-1}(F) = \{r \in V : \vartheta(r) \in F\} \tag{3}$$

for every set F in $\mathfrak{C}$, by Theorem 2.16. The composition

$$\delta = \psi \circ \varphi \circ \varrho^{-1} \tag{4}$$

is a monomorphism from $\mathfrak{C}_0$ to $\mathfrak{B}_0$ (see the diagram below).

$$\mathfrak{B} \xleftarrow{\ \varphi\ } \mathfrak{C} \qquad\qquad \mathfrak{V} \xrightarrow{\ \vartheta\ } \mathfrak{W}$$

$$\psi\downarrow \qquad\quad \downarrow\varrho$$

$$\mathfrak{B}_0 \xleftarrow{\ \delta\ } \mathfrak{C}_0$$

We proceed to show that δ is the identity function on $\mathfrak{C}_0$. From this it follows at once that $\mathfrak{C}_0$ is a subalgebra of $\mathfrak{B}_0$. For each element G in $\mathfrak{C}_0$, there is a unique element F in $\mathfrak{C}$ such that

$$G = \varrho(F) = F \cap U, \tag{5}$$

by (2) and the fact that ϱ is an isomorphism from $\mathfrak{C}$ to $\mathfrak{C}_0$. An easy computation yields

$$\delta(G) = \psi(\varphi(\varrho^{-1}(G))) = \psi(\varphi(F))$$
$$= \psi(\vartheta^{-1}(F)) = \vartheta^{-1}(F) \cap U = F \cap U = G.$$

The first equality follows from (4), the second and final equalities from (5), the third equality from (3), and the fourth equality from (1). For the fifth equality, observe that an element r is in $\vartheta^{-1}(F) \cap U$ just in case r is in U and $\vartheta(r)$ is in F, by (3). Since ϑ is the identity function on U, this last condition is equivalent to saying that r is in $F \cap U$. Conclusion: δ is the identity function on $\mathfrak{C}_0$.

To prove the converse direction of the lemma, assume that $\mathfrak{C}_0$ is a subalgebra of $\mathfrak{B}_0$. For every set F in $\mathfrak{C}$, there is then a unique set G in $\mathfrak{B}$ such that

$$F \cap U = G \cap U. \tag{6}$$

Indeed, for every set F in $\mathfrak{C}$, the intersection $F \cap U$ belongs to $\mathfrak{C}_0$, by the definition of $\mathfrak{C}_0$, and therefore this intersection belongs to $\mathfrak{B}_0$, by the assumption that $\mathfrak{C}_0$ is a subalgebra of $\mathfrak{B}_0$. The mapping ψ is an isomorphism from $\mathfrak{B}$ to $\mathfrak{B}_0$, so there must be a unique set G in $\mathfrak{B}$ that is mapped to $F \cap U$ by ψ. But ψ maps G to $G \cap U$, by (1), so G is the unique set in $\mathfrak{B}$ such that (6) holds.

The identity function on $\mathfrak{C}_0$—call it δ—is a monomorphism from $\mathfrak{C}_0$ into $\mathfrak{B}_0$, by the assumption that $\mathfrak{C}_0$ is a subalgebra of $\mathfrak{B}_0$. The composition

$$\varphi = \psi^{-1} \circ \delta \circ \varrho \tag{7}$$

is therefore a monomorphism from $\mathfrak{C}$ into $\mathfrak{B}$ (see the diagram above). For each set F in $\mathfrak{C}$, if G is the unique set in $\mathfrak{B}$ such that (6) holds, then

$$\varphi(F) = \psi^{-1}(\delta(\varrho(F))) = \psi^{-1}(\delta(F \cap U))$$
$$= \psi^{-1}(F \cap U) = \psi^{-1}(G \cap U) = G,$$

by (7), (2), the assumption that δ is the identity function on $\mathfrak{C}_0$, (6), and (1). Since φ is a monomorphism, and since G is the unique set in $\mathfrak{B}$ for which (6) holds, it may be concluded that

$$\varphi(F) = G \qquad \text{if and only if} \qquad F \cap U = G \cap U \tag{8}$$

for all sets F in $\mathfrak{C}$ and G in $\mathfrak{B}$.

The topological dual of φ is the continuous bounded epimorphism ϑ from $\mathfrak{V}$ to $\mathfrak{W}$ that is determined by

$$\vartheta(u) \in F \qquad \text{if and only if} \qquad u \in \varphi(F) \tag{9}$$

for every set F in $\mathfrak{C}$, by Corollary 2.19 and the remarks following that corollary (with ϑ and $\mathfrak{W}$ in place of δ and $\mathfrak{U}$ respectively). To check that ϑ is the identity function on $\mathfrak{U}$, consider an arbitrary element u in $\mathfrak{U}$. If F is a set in $\mathfrak{C}$ and if $\varphi(F) = G$, then

$$\vartheta(u) \in F \qquad \text{if and only if} \qquad u \in \varphi(F),$$
$$\text{if and only if} \qquad u \in G,$$
$$\text{if and only if} \qquad u \in G \cap U,$$
$$\text{if and only if} \qquad u \in F \cap U,$$
$$\text{if and only if} \qquad u \in F,$$

by (9), the definition of G, the assumption that u is in $\mathfrak{U}$, and (8). These equivalences hold for every set F in $\mathfrak{C}$, that is to say, for every clopen subset F of $\mathfrak{W}$. Since the clopen sets separate points in $\mathfrak{W}$, it follows that $\vartheta(u)$ and u cannot be distinct points, so $\vartheta(u) = u$. Thus, ϑ is the identity function on elements in $\mathfrak{U}$. $\quad\square$

The argument following (8) shows that if φ is a monomorphism from $\mathfrak{C}$ into $\mathfrak{B}$ satisfying condition (8), then the topological dual of φ must be the identity function on the universe of $\mathfrak{U}$. Notice also that the epimorphism involved in the conclusion of the lemma is bounded, and not just weakly bounded.

One consequence of Lemma 3.12 is that two topological extensions of a relational structure $\mathfrak{U}$ are homeo-isomorphic over $\mathfrak{U}$ if and only if the relativizations of their dual algebras to U are equal. The next lemma contains a precise formulation of this remark and is the analogue of Lemma 2.52.

Lemma 3.13. *Let $\mathfrak{V}$ and $\mathfrak{W}$ be topological extensions of a relational structure $\mathfrak{U}$, and $\mathfrak{B}$ and $\mathfrak{C}$ their respective dual algebras. The spaces $\mathfrak{V}$ and $\mathfrak{W}$ are homeo-isomorphic via a function that is the identity function on $\mathfrak{U}$ if and only if the relativizations of $\mathfrak{B}$ and $\mathfrak{C}$ to U are equal, or equivalently, if and only if $\mathfrak{C}$ is isomorphic to $\mathfrak{B}$ via a function φ satisfying the condition*

$$\varphi(F) = G \qquad \text{if and only if} \qquad F \cap U = G \cap U$$

for every F in $\mathfrak{C}$ and G in $\mathfrak{B}$.

Proof. The proof of the first equivalence of the lemma is nearly the same as the corresponding part of the proof of Lemma 2.52 (with references to Lemma 2.51 and Definition 2.48 replaced by references to Lemma 3.12 and Definition 3.11 respectively), so the details are left to the reader. To establish the second equivalence of the lemma, assume that $\mathfrak{C}$ is isomorphic to $\mathfrak{B}$ via a function φ that satisfies the condition stated in the lemma. The topological dual of φ is certainly a homeo-isomorphism from $\mathfrak{W}$ to $\mathfrak{V}$, by Corollary 2.19, and the restriction of this dual to the universe of $\mathfrak{U}$ is the identity function on that universe, by the remark following Lemma 3.12.

Assume now that $\mathfrak{V}$ is homeo-isomorphic to $\mathfrak{W}$ via a function ϑ that is the identity function on $\mathfrak{U}$. The relativizations $\mathfrak{B}_0$ and $\mathfrak{C}_0$ are then equal, by the first part of the lemma. The argument establishing (6) in the proof of Lemma 3.12 shows that for every set F in $\mathfrak{C}$, there is a unique set G in $\mathfrak{B}$ such that

$$F \cap U = G \cap U, \tag{1}$$

and vice versa. If ψ is the relativization isomorphism from $\mathfrak{B}$ to $\mathfrak{B}_0$, and ϱ the relativization isomorphism from $\mathfrak{C}$ to $\mathfrak{C}_0$, then the composition

$$\varphi = \psi^{-1} \circ \varrho$$

is an isomorphism from $\mathfrak{C}$ to $\mathfrak{B}$, since $\mathfrak{B}_0 = \mathfrak{C}_0$. For each set F in $\mathfrak{C}$, if G is the unique set in $\mathfrak{B}$ satisfying (1), then

$$\varphi(F) = \psi^{-1}(\varrho(F)) = \psi^{-1}(F \cap U) = \psi^{-1}(G \cap U) = G,$$

by the definitions of the mappings φ, ψ, and ϱ, together with (1). Consequently, the isomorphism φ satisfies the condition stated at the end of the lemma. $\square$

Ideally, every subalgebra of $\mathfrak{Cm}(U)$ should be realizable as the relativization to U of the dual algebra of some topological extension of $\mathfrak{U}$. Unfortunately, some subalgebras of $\mathfrak{Cm}(U)$ retain so little of the structure of $\mathfrak{U}$ that it seems hopeless to recreate $\mathfrak{U}$ inside their dual spaces. What is true is that every subalgebra of $\mathfrak{Cm}(U)$ can be realized as a relativization of the dual algebra of some relational space $\mathfrak{V}$ to a dense, weakly bounded homomorphic image (as opposed to an isomorphic image) of $\mathfrak{U}$. The following notation will be needed to formulate a precise version of this observation. If ϱ is a function defined on a set U, then ϱ induces a mapping $\bar{\varrho}$ on subsets of U that is defined by

$$\bar{\varrho}(F) = \{\varrho(r) : r \in F\}$$

for each subset F of U.

Lemma 3.14. *Let $\mathfrak{U}$ be a relational structure, and $\mathfrak{C}$ an arbitrary subalgebra of $\mathfrak{Cm}(U)$. If $\mathfrak{V}$ is the dual space of $\mathfrak{C}$, and $\mathfrak{B}$ the dual algebra of $\mathfrak{V}$, then there is a weakly bounded homomorphism ϱ from $\mathfrak{U}$ onto a dense, weak inner substructure of $\mathfrak{V}$ such that the mapping induced on $\mathfrak{C}$ by ϱ maps $\mathfrak{C}$ isomorphically to the relativization of $\mathfrak{B}$ to the universe of that substructure.*

Proof. Let $\mathfrak{C}$ be a subalgebra of $\mathfrak{Cm}(U)$, and take $\mathfrak{V}$ to be the (topological) dual relational space of $\mathfrak{C}$ (see Theorem 2.8 and the remarks following it). The elements in $\mathfrak{V}$ are the ultrafilters in $\mathfrak{C}$, and the topology on $\mathfrak{V}$ is the one induced by $\mathfrak{C}$. Thus, the clopen sets in $\mathfrak{V}$ are the sets of the form
$$H_F = \{Y \in V : F \in Y\}, \tag{1}$$
where F ranges over the elements in $\mathfrak{C}$; and the open sets in $\mathfrak{V}$ are the unions of systems of clopen sets.

For every element r in $\mathfrak{U}$, the set

$$Y_r = \{F \in C : r \in F\} \tag{2}$$

is an ultrafilter in $\mathfrak{C}$ and therefore an element in $\mathfrak{V}$. Define a subset W of $\mathfrak{V}$ by

$$W = \{Y_r : r \in U\}. \tag{3}$$

It is not difficult to check that W is dense in $\mathfrak{V}$. Indeed, an arbitrary non-empty clopen set in $\mathfrak{V}$ has the form (1) for some non-empty set F in $\mathfrak{C}$. From (1)–(3), and the fact that elements in $\mathfrak{C}$ are subsets of $\mathfrak{U}$, it follows that

$$H_F \cap W = \{Y_r : r \in U \text{ and } F \in Y_r\} = \{Y_r : r \in F\}. \tag{4}$$

Because F is assumed to be non-empty, the intersection in (4) is non-empty. Thus, every non-empty clopen set (and hence every non-empty open set) in $\mathfrak{V}$ has a non-empty intersection with W, so W is dense in $\mathfrak{V}$, as claimed.

Let $\mathfrak{W}$ be the restriction of $\mathfrak{V}$ to the set W. We shall prove that $\mathfrak{W}$ is a weakly bounded homomorphic image of $\mathfrak{U}$ under the function ϱ that maps each element r in $\mathfrak{U}$ to the element Y_r in $\mathfrak{W}$. The function ϱ certainly maps the set U onto the set W, by (3) and the definition of ϱ. To check that ϱ is a weakly bounded homomorphism, consider as an example the case of a ternary relation R and the corresponding binary operator $\circ$. Let r, s, and t be arbitrary elements in $\mathfrak{U}$, and suppose

$$R(r, s, t) \tag{5}$$

holds in $\mathfrak{U}$. In this case,

$$t \in \{r\} \circ \{s\} \tag{6}$$

in $\mathfrak{Cm}(U)$, by the definition of the operator $\circ$ in $\mathfrak{Cm}(U)$. If F and G are sets in Y_r and Y_s respectively, then r is in F and s in G, by (2), and therefore

$$\{r\} \circ \{s\} \subseteq F \circ G,$$

by the monotony of the operator $\circ$ in $\mathfrak{Cm}(U)$. Consequently, t belongs to $F \circ G$, by (6), so $F \circ G$ belongs to the ultrafilter Y_t, by (2) (with t in place of r). This argument shows that

$$Y_r \circ Y_s \subseteq Y_t, \tag{7}$$

where the product on the left is the complex product of the subsets Y_r and Y_s of $\mathfrak{Cm}(U)$. It follows from (7) and the definition of the relation R in $\mathfrak{V}$ (the dual space of $\mathfrak{C}$) that

$$R(Y_r, Y_s, Y_t) \tag{8}$$

holds in $\mathfrak{V}$. Since each of the ultrafilters Y_r, Y_s, and Y_t belongs to the set W, by (3), and since $\mathfrak{W}$ is the restriction of $\mathfrak{V}$ to W, we arrive at the conclusion that (8) holds in $\mathfrak{W}$ as well. Thus, (5) implies (8), so ϱ is a homomorphism.

The next step is to prove that ϱ is weakly bounded as a homomorphism. To this end, let Y_1 and Y_2 be elements in $\mathfrak{V}$, and t an element in $\mathfrak{U}$, and consider clopen subsets

$$H_F = \{Y \in V : F \in Y\} \quad \text{and} \quad H_G = \{Y \in V : G \in Y\}$$

of $\mathfrak{V}$ that contain Y_1 and Y_2 respectively (where F and G are elements in $\mathfrak{C}$ and hence subsets of $\mathfrak{U}$). Assume

$$R(Y_1, Y_2, \varrho(t)) \tag{9}$$

holds in $\mathfrak{V}$, with the goal of constructing elements r in $\rho^{-1}(H_F)$ and s in $\rho^{-1}(H_G)$ such that (5) holds in $\mathfrak{U}$. Observe that

$$\varrho^{-1}(H_F) = \{r \in U : \varrho(r) \in H_F\} = \{r \in U : Y_r \in H_F\}$$
$$= \{r \in U : F \in Y_r\} = \{r \in U : r \in F\} = F,$$

by the definition of the inverse image of a set under a function, the definition of the mapping ϱ, (1), (2), and the fact that F is a subset of $\mathfrak{U}$. Similarly,

$$\varrho^{-1}(H_G) = G.$$

Thus, we must construct elements r in F and s in G such that (5) holds in $\mathfrak{U}$. Assumption (9) implies that $R(Y_1, Y_2, Y_t)$ holds in $\mathfrak{V}$, by the definition of $\varrho(t)$, so

$$Y_1 \circ Y_2 \subseteq Y_t,$$

by the definition of the relation R in $\mathfrak{V}$. This inclusion means that if $\bar{F}$ and $\bar{G}$ are any sets in Y_1 and Y_2 respectively, then the product $\bar{F} \circ \bar{G}$ (formed in $\mathfrak{Cm}(U)$ or in $\mathfrak{C}$) belongs to Y_t and therefore contains the element t, by (2). Since Y_1 and Y_2 belong to H_F and H_G respectively, by assumption, the sets F and G do belong to Y_1 and Y_2 respectively, by (1), and therefore t does belong to the product $F \circ G$, by the preceding observations. This means that there must be elements r in F and s in G such that (5) holds in $\mathfrak{U}$, by the definition of the operator $\circ$ in $\mathfrak{Cm}(U)$. Thus, ϱ is weakly bounded as a homomorphism, by Definition 3.1.

One consequence of the observations of the preceding paragraph is that $\mathfrak{W}$ is a weak inner substructure of $\mathfrak{V}$, by Lemma 3.8.

Consider now the topological dual algebra $\mathfrak{B}$ of the relational space $\mathfrak{V}$ (see Lemma 2.9 and the remark following it). The algebra $\mathfrak{B}$ is the second dual of $\mathfrak{C}$, so $\mathfrak{C}$ is isomorphic to $\mathfrak{B}$ via the function φ that maps each set F in $\mathfrak{C}$ to the set H_F in $\mathfrak{B}$, by Theorem 2.10. If $\mathfrak{B}_0$ is the relativization of $\mathfrak{B}$ to the set W, then $\mathfrak{B}$ is isomorphic to $\mathfrak{B}_0$ via the relativization isomorphism ψ that maps each set H_F in $\mathfrak{B}$ to the set $H_F \cap W$, by Lemma 3.10 (with $\mathfrak{W}$ in place of $\mathfrak{U}$, and using the fact that W is dense in $\mathfrak{V}$). The composition $\psi \circ \varphi$ is the isomorphism from $\mathfrak{C}$ to $\mathfrak{B}_0$ that is determined by

$$(\psi \circ \varphi)(F) = \psi(\varphi(F)) = \psi(H_F) = H_F \cap W = \{Y_r : r \in F\}, \qquad (10)$$

by the definitions of φ and ψ, and by (4).

The homomorphism ϱ from $\mathfrak{U}$ onto $\mathfrak{W}$ induces a natural mapping $\bar{\varrho}$ from $\mathfrak{C}$ into $\mathfrak{Cm}(W)$ that is defined by

$$\bar{\varrho}(F) = \{\varrho(r) : r \in F\} = \{Y_r : r \in F\} \qquad (11)$$

for every set F in $\mathfrak{C}$. A comparison of (10) and (11) leads to the conclusion that the isomorphism $\psi \circ \varphi$ is just $\bar{\varrho}$. Thus, $\bar{\varrho}$ maps $\mathfrak{C}$ isomorphically to the algebra $\mathfrak{B}_0$ that is the relativization to W of the dual algebra of $\mathfrak{V}$. $\square$

The final lemma sharpens the preceding result in the case when a subalgebra $\mathfrak{C}$ of $\mathfrak{Cm}(U)$ does retain enough information about the structure of $\mathfrak{U}$ to recreate $\mathfrak{U}$ inside of the dual space of $\mathfrak{C}$. It says that if $\mathfrak{C}$ contains all singleton subsets (and therefore all finite subsets) of $\mathfrak{U}$, then $\mathfrak{C}$ can be realized as the relativization to U of the dual algebra of some topological extension of $\mathfrak{U}$. The lemma is the analogue of Lemma 2.53, and is a kind of converse to Lemma 3.13.

Lemma 3.15. *If $\mathfrak{U}$ is a relational structure, and $\mathfrak{C}$ a subalgebra of $\mathfrak{Cm}(U)$ that contains every singleton subset, then there exists a topological extension $\mathfrak{V}$ of $\mathfrak{U}$ such that the relativization to U of the dual algebra of $\mathfrak{V}$ is just $\mathfrak{C}$. Moreover, $\mathfrak{V}$ has the property that every subset of $\mathfrak{U}$ is open in the topology of $\mathfrak{V}$.*

Proof. We continue with the symbolic notation introduced in the proof of Lemma 3.14. Because $\mathfrak{C}$ contains the singleton subsets of $\mathfrak{U}$, the ultrafilter Y_r, which consists of the sets in $\mathfrak{C}$ that contain the element r, must contain the singleton $\{r\}$, and in fact Y_r must be the principal ultrafilter generated by $\{r\}$. Distinct elements r and s in $\mathfrak{U}$ therefore yield distinct ultrafilters Y_r and Y_s in $\mathfrak{C}$, so the function ϱ that maps each element r in $\mathfrak{U}$ to the ultrafilter Y_r is a bijection from the set U to the set

$$W = \{Y_r : r \in U\}.$$

It was shown in Lemma 3.14 that ϱ is a weakly bounded homomorphism from $\mathfrak{U}$ onto $\mathfrak{W}$. By using in an essential way the assumption that $\mathfrak{C}$ contains all singleton subsets of $\mathfrak{U}$, we can now show that ϱ is actually an isomorphism from $\mathfrak{U}$ to $\mathfrak{W}$. For the proof, consider the case of a ternary relation R correlated with a binary operator $\circ$. Assume

$$R(Y_r, Y_s, Y_t)$$

holds in $\mathfrak{W}$, and therefore also in $\mathfrak{V}$, since $\mathfrak{W}$ is a substructure of $\mathfrak{V}$. In view of the definition of the relation R in the dual space $\mathfrak{V}$ of $\mathfrak{C}$, it follows that

$$Y_r \circ Y_s \subseteq Y_t,$$

where the product on the left is the complex product of the subsets Y_r and Y_s in $\mathfrak{C}$ (or in $\mathfrak{Cm}(U)$, since $\mathfrak{C}$ is a subalgebra of $\mathfrak{Cm}(U)$). The singletons $\{r\}$ and $\{s\}$ belong to the ultrafilters Y_r and Y_s respectively, so the product $\{r\} \circ \{s\}$ (formed in $\mathfrak{C}$ or in $\mathfrak{Cm}(U)$) must belong to the complex product $Y_r \circ Y_s$. Since this complex product is included in Y_t, it follows that $\{r\} \circ \{s\}$ belongs to Y_t, and therefore t belongs to $\{r\} \circ \{s\}$, by the definition of Y_t. This implies that $R(r, s, t)$ holds in $\mathfrak{U}$, by the definition of the operator $\circ$ in the complex algebra $\mathfrak{Cm}(U)$, so ϱ isomorphically preserves the relation R. Conclusion: ϱ is an isomorphism from $\mathfrak{U}$ to $\mathfrak{W}$.

We turn to the task of proving that every subset of $\mathfrak{W}$ is open in the topology on $\mathfrak{V}$. This argument, too, requires the assumption that $\mathfrak{C}$ contains all singleton subsets of $\mathfrak{U}$. Consider any singleton subset $\{r\}$ of $\mathfrak{U}$, and observe that

$$H_{\{r\}} = \{Y \in V : \{r\} \in Y\} = \{Y_r\},$$

by the definition of $H_{\{r\}}$ and the definition of Y_r. The set $H_{\{r\}}$ is clopen in $\mathfrak{V}$, by the definition of the topology on $\mathfrak{V}$, because the singleton $\{r\}$ belongs to $\mathfrak{C}$. The definition of the set W implies that every subset of $\mathfrak{W}$ must have the form

$$\{Y_r : r \in Z\}$$

for some subset Z of $\mathfrak{U}$. Since

$$\{Y_r : r \in Z\} = \bigcup\{\{Y_r\} : r \in Z\} = \bigcup\{H_{\{r\}} : r \in Z\},$$

it follows that every subset of $\mathfrak{W}$ is a union of clopen subsets of $\mathfrak{V}$, and therefore every subset of $\mathfrak{W}$ is open in the topology on $\mathfrak{V}$.

Here is a summary of what has been accomplished so far. A subalgebra $\mathfrak{C}$ of $\mathfrak{Cm}(U)$ that contains the singleton subsets of $\mathfrak{U}$ is given; and $\mathfrak{V}$ is taken to be the topological dual space of $\mathfrak{C}$; and $\mathfrak{B}$ is taken to be the dual algebra of $\mathfrak{V}$ and hence the second dual of $\mathfrak{C}$. The relational structure $\mathfrak{U}$ is mapped isomorphically via the function ϱ to a weak inner substructure $\mathfrak{W}$ of $\mathfrak{V}$ that is dense in $\mathfrak{V}$ and that has the property that each of its subsets is open in $\mathfrak{V}$. The given subalgebra $\mathfrak{C}$ is mapped isomorphically via the function $\bar{\varrho}$ induced by ϱ to a subalgebra $\mathfrak{B}_0$ of $\mathfrak{Cm}(W)$ that is the relativization of $\mathfrak{B}$ to W, by Lemma 3.14. If we use the isomorphism ϱ to identify $\mathfrak{U}$ with $\mathfrak{W}$, and we use the induced isomorphism $\bar{\varrho}$ to identify $\mathfrak{C}$ with $\mathfrak{B}_0$, then we arrive at the desired goal: $\mathfrak{V}$ is a relational space that includes $\mathfrak{U}$ as a dense, weak inner substructure, and the relativization of the dual algebra of $\mathfrak{V}$ to U is just $\mathfrak{C}$. Moreover, every subset of $\mathfrak{U}$ is open in $\mathfrak{V}$.

As in the proof of Lemma 2.53, the technical tool for carrying out this identification is a version of the Exchange Principle that applies to structures such as $\mathfrak{U}$. The elements in $\mathfrak{V}$ that come from $\mathfrak{W}$ are replaced by the corresponding elements from $\mathfrak{U}$ (under the correspondence that is the inverse of ϱ), and the remaining elements of $\mathfrak{V}$ are modified if necessary so that they do not occur in $\mathfrak{U}$. Once this is accomplished, the function ϱ becomes the identity function on $\mathfrak{U}$, and therefore the function $\bar{\varrho}$ becomes the identity function on $\mathfrak{C}$. Consequently, $\mathfrak{C}$ coincides with $\mathfrak{B}_0$, which is the relativization of $\mathfrak{B}$ to W (and $W = U$). $\square$

Let K be the class of topological extensions of a fixed relational structure $\mathfrak{U}$. Define a binary relation $\leq$ on K by stipulating for spaces $\mathfrak{V}$ and $\mathfrak{W}$ in K that $\mathfrak{W} \leq \mathfrak{V}$ if and only if there is a continuous bounded epimorphism from $\mathfrak{V}$ to $\mathfrak{W}$ that is the identity function on (elements in) $\mathfrak{U}$. The relation so defined is easily seen to be a quasi-order in the sense that it is reflexive and transitive. Consequently, the relation induces an equivalence relation $\equiv$ on K that is defined by

$$\mathfrak{V} \equiv \mathfrak{W} \qquad \text{if and only if} \qquad \mathfrak{V} \leq \mathfrak{W} \text{ and } \mathfrak{W} \leq \mathfrak{V},$$

and it induces a partial order on the collection of equivalence classes that is defined by saying that the equivalence class of $\mathfrak{W}$ is less than or equal to the equivalence class of $\mathfrak{V}$ just in case $\mathfrak{W} \leq \mathfrak{V}$. It is easy to check that two spaces $\mathfrak{V}$ and $\mathfrak{W}$ in K are equivalent in this sense just in case there is a homeo-isomorphism from $\mathfrak{V}$ to $\mathfrak{W}$ that is the identity function on $\mathfrak{U}$; the argument is similar to the proof of the corresponding result discussed prior to Theorem 2.54.

Two relational spaces $\mathfrak{V}$ and $\mathfrak{W}$ in K are equivalent if and only if their dual algebras are isomorphic via a function φ determined by

$$\varphi(F) = G \qquad \text{if and only if} \qquad F \cap U = G \cap U$$

for clopen sets F in $\mathfrak{V}$ and $\mathfrak{G}$ in $\mathfrak{W}$ (see the second part of Lemma 3.13). In particular, φ must be the identity function on any clopen sets that happen to be subsets of $\mathfrak{U}$. Furthermore, the relational spaces $\mathfrak{V}$ and $\mathfrak{W}$ are equivalent if and only if the isomorphic copies of their dual algebras that are obtained by relativization to U are equal (see the first part of Lemma 3.13). Consequently, one may, with some justification, speak of the dual algebra of the equivalence class of a space in K, and view this dual algebra as a subalgebra of $\mathfrak{Cm}(U)$.

Let ζ be the function that maps each equivalence class of a space $\mathfrak{V}$ in K to the subalgebra $\mathfrak{B}_0$ of $\mathfrak{Cm}(U)$ that is the relativization to U of the dual algebra $\mathfrak{B}$ of $\mathfrak{V}$. It follows from the remarks in the previous paragraph that ζ is a well-defined one-to-one mapping from the partially ordered structure of equivalence classes of spaces in K into the lattice of subalgebras of $\mathfrak{Cm}(U)$. In fact, ζ is a *strong order monomorphism* in the sense that for spaces $\mathfrak{V}$ and $\mathfrak{W}$ in K with dual algebras $\mathfrak{B}$ and $\mathfrak{C}$ respectively, the equivalence class of $\mathfrak{W}$ is less than or equal to the equivalence class of $\mathfrak{V}$ if and only if $\mathfrak{C}_0$ is a subalgebra of $\mathfrak{B}_0$. The proof is just an application of Lemma 3.12, which asserts that $\mathfrak{W} \leq \mathfrak{V}$ if and only if $\mathfrak{C}_0$ is a subalgebra of $\mathfrak{B}_0$.

If $\mathfrak{V}$ is a relational space in K in which every singleton subset (and hence every finite subset) of $\mathfrak{U}$ is open and therefore clopen, then both the dual algebra $\mathfrak{B}$ and the relativized dual algebra $\mathfrak{B}_0$ contain all singletons, and hence all finite sets, in $\mathfrak{Cm}(U)$, by the definitions of these algebras. Conversely, if $\mathfrak{B}_0$ is an arbitrary subalgebra of $\mathfrak{Cm}(U)$ that contains all singletons, then $\mathfrak{B}_0$ is the relativization to U of the dual algebra of a relational space in K in which every singleton subset of $\mathfrak{U}$ is open, by Lemma 3.15. Consequently, the partially ordered structure of equivalence classes of spaces in K in which the singleton subsets of $\mathfrak{U}$ are open is mapped order-isomorphically by ζ to the lattice of subalgebras of $\mathfrak{Cm}(U)$ that contain all singletons. We summarize the results of this section in the following theorem.

Theorem 3.16. *Let $\mathfrak{U}$ be an arbitrary relational structure and K the class of topological extensions of $\mathfrak{U}$. Equivalent relational spaces in K have, up to isomorphism, the same dual algebra, and that dual algebra is isomorphic via relativization to a subalgebra of $\mathfrak{Cm}(U)$. The function ζ that maps each equivalence class of spaces in K to its relativized dual algebra is a strong order monomorphism from the partially ordered structure of equivalence classes of spaces in K into the lattice of subalgebras of $\mathfrak{Cm}(U)$. The partially ordered structure of equivalence classes of spaces in K in which the singleton subsets of $\mathfrak{U}$ are open is mapped order-isomorphically by ζ onto the lattice of subalgebras of $\mathfrak{Cm}(U)$ that contain all singleton subsets of $\mathfrak{U}$.*

3.3 Duality for Complex Algebras

One consequence of Theorem 3.16 is that for each relational structure $\mathfrak{U}$, there is a maximum topological extension of $\mathfrak{U}$ that is unique up to homeo-isomorphisms over $\mathfrak{U}$, and the (topological) dual algebra of this extension is isomorphic via relativization to $\mathfrak{Cm}(U)$. This extension can be characterized in a way that is strikingly similar to the characterization of Stone-Čech compactifications (see Definition 2.55).

Definition 3.17. A relational space $\mathfrak{V}$ is a *Stone-Čech extension* of a relational structure $\mathfrak{U}$ if $\mathfrak{V}$ is a topological extension of $\mathfrak{U}$ and if every weakly bounded homomorphism from $\mathfrak{V}$ into a relational space $\mathfrak{W}$ can be extended to a continuous bounded homomorphism from $\mathfrak{V}$ into $\mathfrak{W}$. $\square$

Notice in this definition that the extension of each weakly bounded homomorphism ϑ from $\mathfrak{U}$ into $\mathfrak{W}$ is required to be a bounded homomorphism, and not just a weakly bounded homomorphism, from $\mathfrak{V}$ into $\mathfrak{W}$. That is because both $\mathfrak{V}$ and $\mathfrak{W}$ are relational spaces. Notice also that the continuous bounded homomorphism extending ϑ is necessarily unique, since any two continuous mappings from a topological space (in this case $\mathfrak{V}$) to a Hausdorff space (in this case $\mathfrak{W}$) that agree on a dense subset (in this case U) agree on the whole space (see Corollary 2 on p. 315 of [10]).

Lemma 3.18. *If $\mathfrak{V}$ is a Stone-Čech extension of a relational structure $\mathfrak{U}$, then every topological extension of $\mathfrak{U}$ is a continuous bounded* *homomorphic image of $\mathfrak{V}$ via a mapping that is the identity function on $\mathfrak{U}$.*

Proof. The proof is similar to the proof of Lemma 2.56. Suppose $\mathfrak{W}$ is a topological extension of $\mathfrak{U}$. Since $\mathfrak{U}$ is, by definition, a dense, weak inner substructure of $\mathfrak{W}$, the identity function ϑ on $\mathfrak{U}$ is (algebraically) a weakly bounded monomorphism from $\mathfrak{U}$ into $\mathfrak{W}$, by Corollary 3.9. Consequently, there must be a continuous bounded homomorphism δ from the Stone-Čech extension $\mathfrak{V}$ into $\mathfrak{W}$ that agrees with ϑ on $\mathfrak{U}$, by Definition 3.17. It remains to check that δ maps $\mathfrak{V}$ onto $\mathfrak{W}$.

A continuous mapping of a compact space into a Hausdorff space maps closed sets to closed sets (see Corollary 1 on p. 315 of [10]), so the image $\delta(V)$ of the universe of $\mathfrak{V}$ under the mapping δ is a closed subset of $\mathfrak{W}$. This image includes the universe of $\mathfrak{U}$ because δ extends the mapping ϑ, which is the identity function on $\mathfrak{U}$. The closure of U in $\mathfrak{W}$ is therefore included in the closed set $\delta(V)$. But U is a dense subset of $\mathfrak{W}$, by Definition 3.11, so the closure of U in $\mathfrak{W}$ is just the universe of $\mathfrak{W}$. Thus, δ maps $\mathfrak{V}$ onto $\mathfrak{W}$, as required. $\square$

The preceding lemma implies the uniqueness, up to homeo-isomorphisms, of a Stone-Čech extension of a relational structure $\mathfrak{U}$, and justifies the practice of speaking about *the* Stone-Čech extension of $\mathfrak{U}$. The proof of this fact is nearly identical to the proof of Corollary 2.57, and is left to the reader.

Corollary 3.19. *A Stone-Čech extension of a relational structure $\mathfrak{U}$, if it exists, is unique up to homeo-isomorphisms that are the identity on $\mathfrak{U}$.*

We turn now to the question of the existence of Stone-Čech extensions.

Theorem 3.20. *The Stone-Čech extension of an arbitrary relational structure $\mathfrak{U}$ exists, and the dual algebra of this extension is isomorphic via relativization to the complex algebra $\mathfrak{Cm}(U)$.*

Proof. The relational structure $\mathfrak{U}$ has a topological extension $\mathfrak{V}$ with the property that its dual algebra—call it $\mathfrak{B}$—is isomorphic to $\mathfrak{Cm}(U)$ via the relativization function ψ defined by

$$\psi(G) = G \cap U$$

for every set G in $\mathfrak{B}$, by Lemma 3.15. The inverse function ψ^{-1} is the isomorphism from $\mathfrak{Cm}(U)$ to $\mathfrak{B}$ that is defined by

$$\psi^{-1}(F) = G \qquad \text{if and only if} \qquad F = G \cap U \qquad (1)$$

for subsets F of U.

In order to prove that $\mathfrak{V}$ is the Stone-Čech extension of $\mathfrak{U}$, it must be shown that every weakly bounded homomorphism ϑ from $\mathfrak{U}$ into a relational space $\mathfrak{W}$ can be extended to a continuous bounded homomorphism δ from $\mathfrak{V}$ into $\mathfrak{W}$. Consider such a mapping ϑ, and let $\mathfrak{C}$ be the (topological) dual algebra of $\mathfrak{W}$. Theorem 3.2 implies that ϑ has a hybrid dual, namely the homomorphism φ from $\mathfrak{C}$ into $\mathfrak{Cm}(U)$ that is defined by

$$\varphi(H) = \vartheta^{-1}(H) = \{u \in U : \vartheta(u) \in H\} \qquad (2)$$

for every set H in $\mathfrak{C}$. The composition of ψ^{-1} with φ (see the diagram below) is the homomorphism ϱ from $\mathfrak{C}$ into $\mathfrak{B}$ that is determined by

$$\varrho(H) = G \qquad \text{if and only if} \qquad \vartheta^{-1}(H) = G \cap U, \qquad (3)$$

by (1), (2), and the assumption that $\varrho = \psi^{-1} \circ \varphi$. (The fact that the relativization of $\mathfrak{B}$ to U is all of $\mathfrak{Cm}(U)$ is being used here to ensure that the domain of ψ^{-1} is $\mathfrak{Cm}(U)$.)

$$\mathfrak{B} \xrightarrow{\;\psi\;} \mathfrak{Cm}(U) \xleftarrow{\;\varphi\;} \mathfrak{C}$$

The algebra $\mathfrak{C}$ is the dual of the relational space $\mathfrak{W}$, and the algebra $\mathfrak{B}$ is the dual of the relational space $\mathfrak{V}$. Consequently, the homomorphism ϱ from $\mathfrak{C}$ into $\mathfrak{B}$ has a topological dual, namely the continuous bounded homomorphism δ from $\mathfrak{V}$ into $\mathfrak{W}$ that is defined by

$$\delta(s) = r \qquad \text{if and only if} \qquad \varrho^{-1}(Y_s) = X_r,$$

where

$$X_r = \{H \in C : r \in H\} \qquad \text{and} \qquad Y_s = \{G \in B : s \in G\},$$

by Corollary 2.19 (with $\mathfrak{U}$, $\mathfrak{A}$, and φ replaced by $\mathfrak{W}$, $\mathfrak{C}$, and ϱ respectively). It remains to show that δ is an extension of ϑ.

According to the remarks following Corollary 2.19, the function δ is completely determined by the equivalence

$$\delta(s) \in H \qquad \text{if and only if} \qquad s \in \varrho(H) \qquad (4)$$

for all elements s in $\mathfrak{V}$ and all sets H in $\mathfrak{C}$. Fix such a set H, and write $\varrho(H) = G$. For any element s in $\mathfrak{U}$, we have

$$
\begin{aligned}
s \in \varrho(H) \qquad &\text{if and only if} \qquad s \in G, \\
&\text{if and only if} \qquad s \in G \cap U, \\
&\text{if and only if} \qquad s \in \vartheta^{-1}(H), \\
&\text{if and only if} \qquad \vartheta(s) \in H,
\end{aligned}
$$

by the definition of the set G, the assumption that s is in $\mathfrak{U}$, the equivalence in (3), and the definition of the inverse image of the set H under the mapping ϑ. Combine the preceding equivalences with (4) to arrive at

$$\delta(s) \in H \qquad \text{if and only if} \qquad \vartheta(s) \in H. \tag{5}$$

The clopen subsets of $\mathfrak{W}$—that is to say, the sets in $\mathfrak{C}$—separate points, because $\mathfrak{W}$ is a Boolean space. Consequently, (5) can only happen if

$$\delta(s) = \vartheta(s).$$

This is true for every element s in $\mathfrak{U}$, so δ agrees with ϑ on $\mathfrak{U}$, as desired. $\quad\square$

Theorem 3.20 yields the following characterization of the relational space whose dual algebra is isomorphic to $\mathfrak{Cm}(U)$.

Corollary 3.21. *If $\mathfrak{V}$ is a topological extension of a relational structure $\mathfrak{U}$, and $\mathfrak{B}$ the dual algebra of $\mathfrak{V}$, then $\mathfrak{V}$ is the Stone-Čech extension of $\mathfrak{U}$ if and only if the relativization of $\mathfrak{B}$ to U is $\mathfrak{Cm}(U)$.*

Proof. If $\mathfrak{V}$ is the Stone-Čech extension of a relational structure $\mathfrak{U}$, then every topological extension of $\mathfrak{U}$ is a continuous bounded homomorphic image of $\mathfrak{V}$ via a mapping that is the identity function on $\mathfrak{U}$, by Lemma 3.18. This means that every topological extension of $\mathfrak{U}$ is below $\mathfrak{V}$ in the quasi-ordering on the class $\mathbf{K}$ of topological extensions of $\mathfrak{U}$. In other words, the equivalence class of $\mathfrak{V}$ is the maximum element in the partially ordered structure of equivalence classes of spaces in $\mathbf{K}$. Consequently, the relativization of the dual algebra $\mathfrak{B}$ of $\mathfrak{V}$ to the set U is $\mathfrak{Cm}(U)$, by Theorem 3.16.

Conversely, if the relativization of $\mathfrak{B}$ to the set U is $\mathfrak{Cm}(U)$, then the equivalence class of $\mathfrak{V}$ must be the maximum element in the partially ordered structure of equivalence classes of $\mathbf{K}$, again by Theorem 3.16. Consequently, $\mathfrak{V}$ is the Stone-Čech extension of $\mathfrak{U}$ by (the proof of) Theorem 3.20. $\quad\square$

Theorem 3.20 does not automatically imply that the topology on the Stone-Čech extension $\mathfrak{V}$ of a relational structure $\mathfrak{U}$ is in any way related to the classic Stone-Čech compactification of a topological space. Showing that the topology on $\mathfrak{V}$ really is the classic Stone-Čech compactification of some topological space requires a separate argument. (An analogous remark was made in Section 2.12 regarding the notion of a Stone-Čech compactification that was introduced in Definition 2.55.) We shall do this in the next section, but the essential points of the argument are contained in the next lemma and theorem.

Lemma 3.22. *Any two disjoint subsets of a relational structure $\mathfrak{U}$ are separated by a clopen subset of the Stone-Čech extension of $\mathfrak{U}$.*

Proof. Let $\mathfrak{V}$ be the Stone-Čech extension of $\mathfrak{U}$, and $\mathfrak{B}$ the dual algebra of $\mathfrak{V}$. The relativization mapping ψ defined by

$$\psi(F) = F \cap U$$

for sets F in $\mathfrak{B}$ is an isomorphism from $\mathfrak{B}$ to $\mathfrak{Cm}(U)$, by Theorem 3.20. The hybrid dual of the isomorphism ψ is the weakly bounded homomorphism ϑ from $\mathfrak{U}$ into $\mathfrak{V}$ that is defined by

$$\vartheta(u) = r \quad \text{if and only if} \quad r \in \bigcap \{F \in B : u \in \varphi(F)\}$$

for elements u in $\mathfrak{U}$, by Theorem 3.3. Since ψ maps $\mathfrak{B}$ onto $\mathfrak{Cm}(U)$, the inverse images under ϑ of clopen subsets of $\mathfrak{V}$ separate disjoint subsets of $\mathfrak{U}$, by Theorem 3.4. The mapping ϑ is characterized by the equation

$$\vartheta^{-1}(F) = \psi(F)$$

for sets F in $\mathfrak{B}$, by the remarks following Theorem 3.3, so for any two disjoint subsets F_1 and F_2 of $\mathfrak{U}$, there must be a clopen subset G of $\mathfrak{V}$—that is to say, there must be an element G in $\mathfrak{B}$—such that

$$F_1 \subseteq \vartheta^{-1}(G) = \psi(G) = G \cap U, \quad \text{and} \quad \vartheta^{-1}(G) \cap F_2 = \varnothing.$$

It follows from the second equation that

$$F_2 \subseteq {\sim}\vartheta^{-1}(G) = \vartheta^{-1}({\sim}G) = \psi({\sim}G) = U \sim G.$$

Thus, the clopen set G separates the disjoint sets F_1 and F_2. $\square$

Theorem 3.23. *If $\mathfrak{V}$ is the Stone-Čech extension of a relational structure $\mathfrak{U}$, then any function from $\mathfrak{U}$ into a compact Hausdorff space W can be extended to a continuous function from $\mathfrak{V}$ into W.*

Proof. The proof is very similar to the proof of Theorem 2.59. For every element s in $\mathfrak{V}$, define N_s to be the set of clopen subsets of $\mathfrak{V}$ that contain s, and observe that N_s is an ultrafilter in the dual algebra of $\mathfrak{V}$. Define M_s to be the set of sets of the form $\vartheta(F \cap U)^-$ such that F is in N_s. Just as in the proof of Theorem 2.59, one shows that the sets in M_s are not empty and the intersection of every finite system of sets in M_s includes a set from M_s, so that M_s has the finite intersection property; and then one applies compactness to conclude that the intersection of the sets in M_s is not empty. The proof that this intersection cannot contain two elements proceeds as follows.

Consider any two distinct points r_1 and r_2 in W. The goal is to construct a set that belongs to M_s and does not contain one of these two points. It then follows that the intersection of the sets in M_s cannot contain two points. Since W is a Hausdorff space, there are disjoint open sets H_1 and H_2 that contain r_1 and r_2 respectively. The inverse images $\vartheta^{-1}(H_1)$ and $\vartheta^{-1}(H_2)$ are disjoint subsets of $\mathfrak{U}$, so by Lemma 3.22, there must be a clopen set G in $\mathfrak{V}$ that includes $\vartheta^{-1}(H_1)$ and is disjoint from $\vartheta^{-1}(H_2)$. Either G or $\sim G$ is in N_s, since N_s is an ultrafilter. If G belongs to this ultrafilter, then the set $\vartheta(G \cap U)^-$ is in M_s, by the definition of M_s; the point r_2 does not belong to this set and therefore r_2 cannot belong to the intersection of the sets in M_s. If $\sim G$ is in N_s, then the set $\vartheta(U \sim G)^-$ belongs to M_s; the point r_1 does not belong to this set and therefore r_1 cannot belong to the intersection of the sets in M_s.

In more detail, if r_2 were in the closure $\vartheta(G \cap U)^-$, then every open set containing r_2 would have a non-empty intersection with the set $\vartheta(G \cap U)$. In particular, H_2 would have such a non-empty intersection. This implies that there would be a point t in $G \cap U$ such that $\vartheta(t)$ is in H_2. Consequently, the point t would belong to the intersection of G and $\vartheta^{-1}(H_2)$, contradicting the fact that these two sets are disjoint. Similarly, if r_1 were in $\vartheta(U \sim G)^-$, then every open set containing r_1 would have a non-empty intersection with $\vartheta(U \sim G)$. In particular, H_1 would have such a non-empty intersection. This implies that there would be a point t in $U \sim G$ such that $\vartheta(t)$ is in H_1. In other words, the point t would be in $\sim G$ and in $\vartheta^{-1}(H_1)$, contradicting the fact that this last set is included in G.

Define the function δ on each point s in $\mathfrak{V}$ by taking $\delta(s)$ to be the unique point that belongs to the intersection of M_s. The arguments showing that the function δ is continuous and that δ agrees with ϑ

on U are virtually identical to the arguments in the corresponding parts of the proof of Theorem 2.59, and are left to the reader. □

The compact Hausdorff space W in Theorem 3.23 may be the universe of a relational structure $\mathfrak{W}$, and the given function ϑ from $\mathfrak{U}$ into W may be a homomorphism (not necessarily weakly bounded) with respect to the fundamental relations of $\mathfrak{V}$ and $\mathfrak{W}$. If $\mathfrak{W}$ is a *closed space* in the sense that every relation in $\mathfrak{W}$ of rank n is a closed subset of the product space W^n (see the remarks preceding Theorem 2.60), then the (unique) extension of ϑ to a continuous mapping from $\mathfrak{V}$ to W will also be a homomorphism, as the next theorem shows. The theorem is an analogue of Theorem 2.60.

Theorem 3.24. *If $\mathfrak{V}$ is the Stone-Čech extension of a relational structure $\mathfrak{U}$, then every homomorphism from $\mathfrak{U}$ into a closed space $\mathfrak{W}$ can be extended to a continuous homomorphism from $\mathfrak{V}$ into $\mathfrak{W}$.*

Proof. Suppose ϑ is a homomorphism from $\mathfrak{U}$ into a closed space $\mathfrak{W}$. The topology on $\mathfrak{W}$ is, by definition, compact and Hausdorff, so there is a uniquely determined continuous function δ from $\mathfrak{V}$ into $\mathfrak{W}$ that extends ϑ, by Theorem 3.23. It must be shown that δ preserves the relations in $\mathfrak{V}$. Focus on the case of a ternary relation R. The strategy employed is very similar to the one employed in the proof of Theorem 2.60.

Assume u, v, and w are elements in $\mathfrak{V}$ such that

$$R(u, v, w) \tag{1}$$

holds in $\mathfrak{V}$, and write

$$r = \delta(u), \qquad s = \delta(v), \qquad t = \delta(w). \tag{2}$$

It must be shown that $R(r, s, t)$ holds in $\mathfrak{W}$. Consider arbitrary open subsets H_1, H_2, and H_3 of $\mathfrak{W}$ that contain the points r, s, and t respectively. We shall prove that the product

$$H_1 \times H_2 \times H_3 \tag{3}$$

always has a non-empty intersection with the relation R in $\mathfrak{W}$. It then follows that the triple (r, s, t) belongs to the closure of the relation R in the product space $W \times W \times W$. Since the relation R is closed, by the assumption that $\mathfrak{W}$ is a closed space, we arrive at the desired conclusion that (r, s, t) belongs to R.

The inverse images $\delta^{-1}(H_1)$ and $\delta^{-1}(H_2)$ are open subsets of $\mathfrak{V}$, because δ is a continuous function, and these inverse images contain the points u and v respectively, by (2) and the assumption that r is in H_1 and s in H_2. The clopen sets form a base for the (Boolean space) topology on $\mathfrak{V}$, so there must be clopen subsets F and G of $\mathfrak{V}$ such that

$$u \in F \text{ and } F \subseteq \delta^{-1}(H_1), \qquad \text{and} \qquad v \in G \text{ and } G \subseteq \delta^{-1}(H_2). \quad (4)$$

The image set

$$R^*(F \times G) = \{z \in V : R(x, y, z) \text{ for some } x \in F \text{ and } y \in G\} \quad (5)$$

is clopen in $\mathfrak{V}$, because $\mathfrak{V}$ is a relational space; and the point w belongs to this set by (1) and (4). The inverse image $\delta^{-1}(H_3)$ is open, by the continuity of δ, and it contains the point w, by the final equation in (2) and the assumption that t is in H_3. Consequently, the intersection

$$R^*(F \times G) \cap \delta^{-1}(H_3) \quad (6)$$

is an open set in $\mathfrak{V}$ that contains the point w. The universe of $\mathfrak{U}$ is dense in $\mathfrak{V}$, by the definition of a topological extension, so the set in (6) must contain a point $\bar{w}$ in $\mathfrak{U}$. In particular, $\bar{w}$ belongs to $R^*(F \times G)$, so there must be points $\bar{u}$ in F and $\bar{v}$ in G such that

$$R(\bar{u}, \bar{v}, \bar{w}) \quad (7)$$

holds in $\mathfrak{V}$, by (5).

Up to this point, the argument has been the same as the corresponding argument in the proof of Theorem 2.60. The relational structure $\mathfrak{U}$ is, however, not an inner substructure of $\mathfrak{V}$, but only a weak inner substructure of $\mathfrak{V}$, by the definition of a topological extension. Consequently, the next part of the argument is necessarily different. Since $\mathfrak{U}$ is a weak inner substructure of $\mathfrak{V}$, and (7) holds in $\mathfrak{V}$, and $\bar{u}$ is in F, and $\bar{v}$ in G, and $\bar{w}$ in U, there must be elements $\hat{u}$ in $F \cap U$ and $\hat{v}$ in $G \cap U$ such that $R(\hat{u}, \hat{v}, \bar{w})$ holds in $\mathfrak{U}$, by Definition 3.6. The function ϑ is a homomorphism from $\mathfrak{U}$ into $\mathfrak{W}$, by assumption, so

$$R(\vartheta(\hat{u}), \vartheta(\hat{v}), \vartheta(\bar{w}))$$

holds in $\mathfrak{W}$. The continuous mapping δ is an extension of ϑ, so

$$R(\delta(\hat{u}), \delta(\hat{v}), \delta(\bar{w})) \quad (8)$$

holds in $\mathfrak{W}$. The point $\hat{u}$ belongs to F, which is included in $\delta^{-1}(H_1)$, by (4), so the image point $\delta(\hat{u})$ is in H_1. Similarly, the point $\hat{v}$ belongs to G, which is included in $\delta^{-1}(H_2)$, so $\delta(\hat{v})$ is in H_2. Finally, $\bar{w}$ is in $\delta^{-1}(H_3)$, by the choice of $\bar{w}$ after (6), so $\delta(\bar{w})$ is in H_3. Thus, the triple $(\delta(\hat{u}), \delta(\hat{v}), \delta(\bar{w}))$ belongs to the product in (3), and it also belongs to the relation R in $\mathfrak{W}$, by (8). Consequently, the product in (3) has a non-empty intersection with the relation R in $\mathfrak{W}$, as was to be shown. $\square$

3.4 Weak Compactifications of Discrete Spaces

Some (but not all) of the results in the preceding two sections can be recast in a topological form that relates subalgebras of the complex algebra $\mathfrak{Cm}(U)$ of a relational structure $\mathfrak{U}$ to the (topological) dual algebras of certain kinds of compactifications of the discretely topologized structure $\mathfrak{U}$. For example, it follows from Theorem 3.16 that for a discretely topologized relational structure $\mathfrak{U}$, every subalgebra of $\mathfrak{Cm}(U)$ that contains all of the singleton subsets of $\mathfrak{U}$ is isomorphic via relativization to the dual algebra of a weak compactification of $\mathfrak{U}$, and vice versa. The function that maps each equivalence class of weak compactifications of $\mathfrak{U}$ to the isomorphic copy of the dual algebra of the equivalence class is an isomorphism from the lattice of equivalence classes of weak compactifications of $\mathfrak{U}$ onto the lattice of subalgebras of $\mathfrak{Cm}(U)$ that contain the singleton subsets. In particular, there is a maximum weak compactification of the discretely topologized space $\mathfrak{U}$, and the dual algebra of this maximum weak compactification is isomorphic via relativization to $\mathfrak{Cm}(U)$. The maximum weak compactification of $\mathfrak{U}$ can be characterized as the Stone-Čech weak compactification of $\mathfrak{U}$. Said somewhat differently, the dual relational space of the complex algebra $\mathfrak{Cm}(U)$ is, up to isomorphism, the Stone-Čech weak compactification of the discretely topologized space $\mathfrak{U}$. We elaborate on these remarks below.

A topology on a set U is said to be *discrete* if every subset of U is open, and therefore every subset of U is closed and hence clopen. A set endowed with the discrete topology is called a *discrete space*. Such a space is always Hausdorff (distinct points r and s are separated by the clopen sets $\{r\}$ and $\{s\}$) and locally compact (every point r belongs to the interior of a compact set, namely $\{r\}$), and the clopen sets form a base for the topology (in fact, they are the topology). A discrete space can only be compact if it is finite, since the singleton subsets form an open cover of the space.

Every relational structure $\mathfrak{U}$, when endowed with the discrete topology, automatically becomes a locally compact relational space in the sense described before Definition 2.48: it is a locally compact Hausdorff space in which the clopen sets form a base for the topology; the relations R of rank $n+1$ in $\mathfrak{U}$ are clopen because the image sets $R^*(F_0 \times \cdots \times F_{n-1})$ are clopen in the topology of $\mathfrak{U}$ for any subsets $F_0, \ldots, F_{n-1}$ of $\mathfrak{U}$ (all subsets of $\mathfrak{U}$ are clopen); and the relations R of rank $n+1 \geq 2$ are continuous because the inverse images under R of (clopen) subsets of $\mathfrak{U}$ are clopen in the product topologies on U^n (every subset of U^n is clopen in the product topology). We shall call such a structure $\mathfrak{U}$ a *discrete relational space*.

Discrete relational spaces are not required to be compact; in fact, they are compact if and only if they are finite. Compactifications of infinite discrete spaces, that is to say, (compact) relational spaces that include $\mathfrak{U}$ as a dense subspace and an inner substructure (see Definition 2.48), exist only in special circumstances, namely when discrete space is of finitary character in the sense defined before Lemma 3.7.

Lemma 3.25. *If a discrete relational space $\mathfrak{U}$ has a compactification, then $\mathfrak{U}$ must be of finitary character.*

Proof. Suppose $\mathfrak{V}$ is a compactification of a discrete space $\mathfrak{U}$, and consider an element t in $\mathfrak{U}$. Focus on the case of a ternary relation R in $\mathfrak{V}$. The set

$$H_t = \{(r, s) \in V \times V : R(r, s, t)\}$$

is closed, and hence compact, in the product space $V \times V$, by Theorem 2.15. Since $\mathfrak{U}$ is an inner substructure of $\mathfrak{V}$, by the definition of a compactification, and since t belongs to $\mathfrak{U}$, it follows that for elements r and s in $\mathfrak{V}$, the relationship $R(r, s, t)$ holds in $\mathfrak{V}$ if and only if r and s both belong to $\mathfrak{U}$ and $R(r, s, t)$ holds in $\mathfrak{U}$ (see Definition 1.12). Consequently,

$$H_t = \{(r, s) \in U \times U : R(r, s, t)\}.$$

In particular, H_t is a subset of $U \times U$ that is compact in $V \times V$, so H_t must remain compact in the subspace $U \times U$ (see Exercise 28 on p. 280 of [10]). Compact subsets of discrete spaces are necessarily finite, so H_t is finite. Conclusion: $\mathfrak{U}$ is of finitary character. $\square$

We shall see in a moment that the converse to the lemma is also true: a discrete relational space that is of finitary character always has compactifications, and in fact it has a Stone-Čech compactification.

The lemma implies that if a relational structure $\mathfrak{U}$ is not of finitary character, then the relational space that is the (topological) dual of the complex algebra $\mathfrak{Cm}(U)$ cannot be a compactification of $\mathfrak{U}$ endowed with the discrete topology, much less the Stone-Čech compactification of $\mathfrak{U}$. As the proof of the lemma demonstrates, the difficulty lies in the requirement that $\mathfrak{U}$ must be an inner substructure of a compactification $\mathfrak{V}$. This requirement is too strong; it needs to be replaced by the weaker requirement that $\mathfrak{U}$ be a weak inner substructure of $\mathfrak{V}$.

Definition 3.26. A (compact) relational space $\mathfrak{V}$ is called a *weak compactification* of a locally compact relational space $\mathfrak{U}$ if $\mathfrak{U}$ is a dense, weak inner subspace of $\mathfrak{V}$ in the following sense: algebraically, $\mathfrak{U}$ is a weak inner substructure of $\mathfrak{V}$; the topology on $\mathfrak{U}$ is the one inherited from $\mathfrak{V}$; and the (topological) closure of the set U in $\mathfrak{V}$ is just the set V. $\square$

The only difference between the notion of a compactification and the notion of a weak compactification of a locally compact relational space $\mathfrak{U}$ is that in the former $\mathfrak{U}$ is required to be an inner substructure (see Definition 2.48), while in the latter $\mathfrak{U}$ is only required to be a weak inner substructure. As the two definitions and Lemmas 3.7 and 3.25 make clear, every compactification of a locally compact relational space $\mathfrak{U}$ is a weak compactification of $\mathfrak{U}$, and if a compactification of $\mathfrak{U}$ exists, then $\mathfrak{U}$ must be of finitary character; and when $\mathfrak{U}$ is of finitary character, the converse is true as well: every weak compactification of $\mathfrak{U}$ is a compactification of $\mathfrak{U}$, because in this case every weak inner substructure is an inner substructure.

The results in the previous two sections can be directly applied to discrete relational spaces. For example, Lemma 3.10 implies that if a relational space $\mathfrak{V}$ is a weak compactification of a discrete relational space $\mathfrak{U}$, and if $\mathfrak{B}$ is the dual algebra of $\mathfrak{V}$, then the set

$$B_0 = \{F \cap U : F \in B\}$$

is a subuniverse of $\mathfrak{Cm}(U)$, and the relativization function mapping F to $F \cap U$ for each set F in $\mathfrak{B}$ is an isomorphism from $\mathfrak{B}$ to the corresponding subalgebra $\mathfrak{B}_0$. In the case of a discrete relational space, one can actually say a bit more; namely, the subalgebra $\mathfrak{B}_0$ contains every singleton subset of $\mathfrak{U}$. The reason is that every dense, locally compact subspace of a Hausdorff space is necessarily an open subset of the Hausdorff space (see Corollary 1 on p. 400 of [10]). In particular, the universe of $\mathfrak{U}$ is an open subset of $\mathfrak{V}$. Since every single subset of $\mathfrak{U}$ is open in $\mathfrak{U}$ (because the topology on $\mathfrak{U}$ is discrete), it follows that every singleton subset of $\mathfrak{U}$ is open in $\mathfrak{V}$. Every singleton subset of $\mathfrak{U}$ is clearly closed in $\mathfrak{V}$, because $\mathfrak{V}$ is a Hausdorff space, so every singleton subset of $\mathfrak{U}$ is clopen in the topology of $\mathfrak{V}$ and is therefore an element in the dual algebra $\mathfrak{B}$. The intersection of such a singleton with the set U is, of course, just the singleton, so every singleton subset of $\mathfrak{U}$ belongs to $\mathfrak{B}_0$, by the definition of $\mathfrak{B}_0$.

An application of Theorem 3.16 to discrete relational spaces leads to the following conclusion.

Theorem 3.27. *Let $\mathfrak{U}$ be an arbitrary discrete relational space, and K the class of weak compactifications of $\mathfrak{U}$. Equivalent relational spaces in K have, up to isomorphism, the same dual algebra, and that dual algebra is isomorphic via relativization to a subalgebra of $\mathfrak{Cm}(U)$ that contains all singleton subsets of $\mathfrak{U}$. The function that maps each equivalence class of weak compactifications in K to its relativized dual algebra is a lattice isomorphism from the lattice of equivalence classes of weak compactifications in K to the lattice of subalgebras of $\mathfrak{Cm}(U)$ that contain all the singleton subsets of $\mathfrak{U}$.*

For discrete relational spaces of finitary character, the concept of a weak compactification coincides with that of a compactification, as was remarked after Definition 3.26. Consequently, if the discrete relational space $\mathfrak{U}$ in the statement of the preceding theorem is of finitary character, then the term "weak compactification" in the theorem may be replaced by the term "compactification".

A set U may be thought of as a relational structure in which there are no fundamental relations. Viewed in this way, U vacuously has finitary character because there are no fundamental relations that do not satisfy the finitary character condition. If U is endowed with the discrete topology, the result is a discrete relational space of finitary character. A compactification of this space is just a Boolean space V that includes U (endowed with the discrete topology) as a dense subspace. Two such compactifications of U are equivalent if there is a homeomorphism from one to the other that is the identity function on U. The complex algebra of U is just the Boolean set algebra $Sb(U)$ of all subsets of U, and the smallest subalgebra of $Sb(U)$ that contains all the singleton subsets of U is just the Boolean set algebra $Cf(U)$ consisting of the finite and the cofinite subsets of U. An application of Theorem 3.27 to U yields the following conclusion.

Corollary 3.28. *Let U be an arbitrary set endowed with the discrete topology, and* K *the class of compactifications of U. Equivalent spaces in* K *have, up to isomorphism, the same dual Boolean algebra, and that dual algebra is isomorphic via relativization to a subalgebra of $Sb(U)$ that includes $Cf(U)$. The function that maps each equivalence class of compactifications in* K *to its relativized dual algebra is a lattice isomorphism from the lattice of equivalence classes of compactifications in* K *to the lattice of subalgebras of $Sb(U)$ that include $Cf(U)$.*

Theorem 3.16 implies that every discrete relational space $\mathfrak{U}$ has, in general, many weak compactifications (and every such space that is of finitary character has many compactifications). In particular, $\mathfrak{U}$ has a maximum weak compactification whose dual algebra is isomorphic to $\mathfrak{Cm}(U)$ via relativization. That maximum weak compactification has all the requisite properties of a Stone-Čech weak compactification.

Definition 3.29. A relational space $\mathfrak{V}$ is a *Stone-Čech weak compactification* of a discrete relational space $\mathfrak{U}$ if $\mathfrak{V}$ is a weak compactification of $\mathfrak{U}$ and if every weakly bounded homomorphism from $\mathfrak{U}$ into a relational space $\mathfrak{W}$ can be extended to a continuous bounded homomorphism from $\mathfrak{V}$ into $\mathfrak{W}$.   □

Every mapping from a discrete relational space $\mathfrak{U}$ into a relational space $\mathfrak{W}$ is automatically continuous, because the inverse image of a subset of $\mathfrak{W}$ under an arbitrary mapping is obviously an open subset of $\mathfrak{U}$ (under the discrete topology on $\mathfrak{U}$). Consequently, it is unnecessary in the preceding definition to require the weakly bounded homomorphism from $\mathfrak{U}$ to $\mathfrak{W}$ to be continuous. For the same reason, the results of Section 3.4—and, in particular, Corollary 3.19 and Theorems 3.20, 3.23, and 3.24—can be directly applied to discrete relational spaces and lead to the following conclusions.

Theorem 3.30. *The Stone-Čech weak compactification $\mathfrak{V}$ of a discrete relational space $\mathfrak{U}$ exists and is unique up to homeo-isomorphisms over $\mathfrak{U}$. The dual algebra of $\mathfrak{V}$ is isomorphic via relativization to the complex algebra $\mathfrak{Cm}(U)$. Every function from $\mathfrak{U}$ into a compact Hausdorff space W can be extended to a continuous function from $\mathfrak{V}$ into W, and every homomorphism from $\mathfrak{U}$ into a closed space $\mathfrak{W}$ can be extended to a continuous homomorphism from $\mathfrak{V}$ into $\mathfrak{W}$.*

As before, if the space $\mathfrak{U}$ in the statement of the preceding theorem is of finitary character, then the term "weak compactification" in the theorem may be replaced by the term "compactification". In particular, in its application to sets (viewed as relational structures with no fundamental relations) endowed with the discrete topology, the preceding theorem implies the well-known result that the Boolean algebra of clopen subsets of the Stone-Čech compactification of a discrete space U is isomorphic to the Boolean algebra of all subsets of U (see Exercise 19 on p. 337 of [10]).

Corollary 3.31. *The Stone-Čech compactification V (in the classic topological sense) of a discrete space U exists and is unique up to homeomorphisms that are the identity function on U. The dual Boolean algebra of V, consisting of the clopen subsets of V, is isomorphic via relativization to the Boolean set algebra $Sb(U)$.*

References

1. Blackburn, P., de Rijke, M., Venema, Y.: Modal Logic. Cambridge Tracts in Theoretical Computer Science, vol. 53, xxii + 554 pp. Cambridge University Press, Cambridge (2001)

2. Celani, S.A.: Topological duality for Boolean algebras with a normal n-ary monotonic operator. Order **26**, 49–67 (2009)

3. Chin, L.H., Tarski, A.: Distributive and modular laws in the arithmetic of relation algebras. University of California Publications in Mathematics, New Series, vol. 1, pp. 341–384. University of California Press, Berkeley/Los Angeles (1951)

4. Everett, C.J., Ulam, S.: Projective algebra I. Am. J. Math. **68**, 77–88 (1946)

5. Feferman, S.: Persistent and invariant formulas for outer extensions. Compos. Math. **20**, 29–52 (1968)

6. Frayne, T., Morel, A.C., Scott, D.S.: Reduced direct products. Fundamenta Mathematicae **51**, 195–228 (1962)

7. Gehrke, M.: Stone duality, topological algebra, and recognition. http://hal.archives-ouvertes.fr/docs/00/86/01/62/PDF/Ge13.pdf

8. Gehrke, M., Harding, J., Venema, Y.: MacNeille completions and canonical extensions. Trans. Am. Math. Soc. **358**, 573–590 (2006)

9. Givant, S.: The Structure of Relation Algebras Generated by Relativizations. Contemporary Mathematics, vol. 156, xvi + 134 pp. American Mathematical Society, Providence (1994)

10. Givant, S., Halmos, P.: Introduction to Boolean Algebras. Undergraduate Texts in Mathematics, xiv + 574 pp. Springer, New York (2009)

11. Goldblatt, R.I.: Metamathematics of Modal Logic. Doctoral dissertation, Victoria University, Wellington, iii + 95 pp. (1974)

12. Goldblatt, R.I.: Metamathematics of modal logic, parts I and II. Rep. Math. Log. **6** and **7**, pp. 41–78 and 21–52 (1976) respectively. (Reprinted in: Goldblatt, R.: Mathematics of Modality. Lecture Notes, vol. 43, pp. 9–79. CSLI Publications, Stanford (1993))

13. Goldblatt, R.I.: Varieties of complex algebras. Ann. Pure Appl. Log. **44**, 173–242 (1989)

14. Goldblatt, R.I.: Maps and monads for modal frames. Stud. Log. **83**, 307–329 (2006)

15. Halmos, P.R.: Algebraic logic I. Monadic Boolean algebras. Compos. Math. **12**, 217–249 (1955)

16. Hansoul, G.: A duality for Boolean algebras with operators. Algebra Universalis **17**, 34–49 (1983)

17. Henkin, L., Tarski, A.: Cylindric algebras. In: Dilworth, R.P. (ed.) Lattice Theory. Proceedings of Symposia in Pure Mathematics, vol. 2, pp. 83–113. American Mathematical Society, Providence (1961)

18. Henkin, L., Monk, J.D., Tarski, A.: Cylindric Algebras, Part I. Studies in Logic and the Foundations of Mathematics, vol. 64, vi + 508 pp. North-Holland, Amsterdam (1971)

19. Jónsson, B.: A survey of Boolean algebras with operators. In: Rosenberg, I.G., Sabidussi, G. (eds.) Algebras and Orders, pp. 239–286. Kluwer Academic, Dordrecht (1993)

20. Jónsson, B., Tarski, A.: Direct Decompositions of Finite Algebraic Systems. Notre Dame Mathematical Lectures, vol. 54, vi + 64 pp. University of Notre Dame, Notre Dame (1947)

21. Jónsson, B., Tarski, A.: Boolean algebras with operators. Part I. Am. J. Math. **73**, 891–939 (1951)

22. Jónsson, B., Tarski, A.: Boolean algebras with operators. Part II. Am. J. Math. **74**, 127–162 (1952)

23. Koppelberg, S.: Handbook of Boolean Algebras, vol. I, xix + 312 pp. North-Holland, Amsterdam (1989)

24. Lemmon, E.J.: Algebraic semantics for modal logics II. J. Symb. Log. **31**, 191–218 (1966)

25. Łoś, J.: Quelques remarques, théorèmes et problèmes sur les classes définissables d'algèbres. In: Mathematical Interpretation of Formal Systems. Studies in Logic and the Foundations of Mathematics, pp. 98–113. North-Holland, Amsterdam (1955)

26. Monk, J.D.: On representable relation algebras. Mich. Math. J. **11**, 207–210 (1964)

27. Monk, J.D.: Completions of Boolean algebras with operators. Math. Nachr. **46**, 47–55 (1970)

28. Peirce, C.S.: Note B. The logic of relatives. In: Peirce, C.S. (ed.) Studies in Logic by Members of the Johns Hopkins University. Little, Brown, and Company, Boston, 1883, pp. 187–203. (Reprinted by John Benjamins Publishing Company, Amsterdam, 1983)

29. Pierce, R.S.: Compact Zero-Dimensional Metric Spaces of Finite Type. Memoirs of the American Mathematical Society, vol. 130, 64 pp. American Mathematical Society, Providence (1972)

30. Priestley, H.A.: Representation of distributive lattices by means of ordered Stone spaces. Bull. Lond. Math. Soc. **2**, 186–190 (1970)

31. Sain, I.: Strong amalgamation and epimorphisms of cylindric algebras and Boolean algebras with operators. Mathematical Institute of the Hungarian Academy of Sciences, 44 pp. Preprint 17/1982 (1982)

32. Sambin, G., Vaccaro, V.: Topology and duality in modal logic. Ann. Pure Appl. Log. **37**, 249–296 (1988)

33. Schröder, E.: Vorlesungen über die Algebra der Logik (exakte Logik), vol. III, Algebra und Logik der Relative, part 1. B. G. Teubner, Leipzig, 1895. (Reprinted by Chelsea Publishing Company, New York, 1966)

34. Segerberg, K.: An Essay in Classical Modal Logic. Filosofiska Studier Utgivna av Filosofiska Föreningen Och Filosofiska Institutionen Vid Uppsala Universitet, vol. 13, 500 pp. University of Uppsala, Uppsala (1971)

35. Sikorski, R.: Boolean Algebras. Ergebnisse der Mathematik und Ihrer Grenzgebiete, New Series, vol. 25. Springer, Berlin/Göttingen/Heidelberg (1960). 3rd edn. (1969)

36. Stone, M.H.: The theory of representations for Boolean algebras. Trans. Am. Math. Soc. **40**, 37–111 (1936)

37. Stone, M.H.: Applications of the theory of Boolean rings to general topology. Trans. Am. Math. Soc. **40**, 321–364 (1937)

38. Tarski, A.: On the calculus of relations. J. Symb. Log. **6**, 73–89 (1941)

39. Thomason, S.K.: Categories of frames for modal logic. J. Symb. Log. **40**, 439–442 (1975)

40. Venema, Y.: Algebras and coalgebras. In: van Benthem, J., Blackburn, P., Wolter, F. (eds.) Handbook of Modal Logic. Studies in Logic and Practical Reasoning, vol. 3, pp. 331–426. Elsevier, Amsterdam (2006)

Index

S. Givant, *Duality Theories for Boolean Algebras with Operators*,
Springer Monographs in Mathematics, DOI 10.1007/978-3-319-06743-8,
© Springer International Publishing Switzerland 2014

Printed in the United States
By Bookmasters